Springer Theses

Recognizing Outstanding Ph.D. Research

For further volumes:
http://www.springer.com/series/8790

Aims and Scope

The series "Springer Theses" brings together a selection of the very best Ph.D. theses from around the world and across the physical sciences. Nominated and endorsed by two recognized specialists, each published volume has been selected for its scientific excellence and the high impact of its contents for the pertinent field of research. For greater accessibility to non-specialists, the published versions include an extended introduction, as well as a foreword by the student's supervisor explaining the special relevance of the work for the field. As a whole, the series will provide a valuable resource both for newcomers to the research fields described, and for other scientists seeking detailed background information on special questions. Finally, it provides an accredited documentation of the valuable contributions made by today's younger generation of scientists.

Theses are accepted into the series by invited nomination only and must fulfill all of the following criteria

- They must be written in good English.
- The topic should fall within the confines of Chemistry, Physics and related interdisciplinary fields such as Materials, Nanoscience, Chemical Engineering, Complex Systems and Biophysics.
- The work reported in the thesis must represent a significant scientific advance.
- If the thesis includes previously published material, permission to reproduce this must be gained from the respective copyright holder.
- They must have been examined and passed during the 12 months prior to nomination.
- Each thesis should include a foreword by the supervisor outlining the significance of its content.
- The theses should have a clearly defined structure including an introduction accessible to scientists not expert in that particular field.

Jamal Jokar Arsanjani

Dynamic Land-Use/Cover Change Simulation: Geosimulation and Multi Agent-Based Modelling

Doctoral Thesis accepted by
University of Vienna, Austria

Springer

Author
Dr. Jamal Jokar Arsanjani
Department of Geography
and Regional Research
University of Vienna
Universitätsstraße 7
A-1010 Vienna
Austria
e-mail: jamaljokar@gmail.com

Supervisor
Prof. Dr. Wolfgang Kainz
Department of Geography
and Regional Research
University of Vienna
Universitätsstraße 7
A-1010 Vienna
Austria

ISSN 2190-5053
e-ISSN 2190-5061
ISBN 978-3-642-27068-0
ISBN 978-3-642-23705-8 (eBook)
DOI 10.1007/978-3-642-23705-8
Springer Heidelberg Dordrecht London New York

Softcover reprint of the hardcover 1st edition 2012

Cover design: eStudio Calamar, Berlin/Figueres

Printed on acid-free paper

Springer is part of Springer Science+Business Media (www.springer.com)

Parts of this thesis have been published in the following journal articles:

J. Jokar Arsanjani, W. Kainz, **Integration Of Spatial Agents And Markov Chain Model in Simulation of Urban Sprawl**, In *Proceeding of AGILE conference 2011*, Utrecht, the Netherlands (*peer reviewed*)

J. Jokar Arsanjani, M. Helbich, W. Kainz, A. Darvishi B., **Integration of Logistic Regression and Markov Chain Models to Simulate Urban Expansion**, Submitted to *the International Journal of Applied Earth Observation and Geoinformation*, 2011 (Accepted for publication)

J. Jokar Arsanjani, W. Kainz, A. Mousivand, **Tracking Dynamic Land Use Change Using Spatially Explicit Markov Chain Based on Cellular Automata-the Case of Tehran**, *International Journal of Image and Data Fusion*, 2011 (In press)

J. Jokar Arsanjani, W. Kainz, M. Azadbakht, **Monitoring and Geospatially Explicit Simulation of Land Use Dynamics: from Cellular Automata towards Geosimulation—Case Study Tehran, Iran**, In *Proceeding of ISDIF 2011*, China

J. Jokar Arsanjani, M. Helbich, W. Kainz, **The Emergence of Urban Sprawl Patterns in Tehran Metropolis through Agent Based Modelling**, in preparation

Supervisor's Foreword

Land use and land cover change are two subjects that have triggered a large number of research activities and resulted in a wealth of different approaches to detect past change and also to predict future development. Among the most prominent methods are those that use remote sensing and image analysis combined with various statistical and analytical procedures. They all require a series of data over longer periods, appropriate land use maps, and related information. It is not always easy to acquire or access these data due to a simple lack of data or administrative access restrictions. It is therefore imperative to make use of satellite data and other easier accessible data of reasonable resolution.

Many large cities face pressing problems with—sometimes uncontrolled—growth and sprawl, in particular when their expansion is limited by natural and other conditions. Tehran is one of these cities whose expansion is a fact, but which also experiences severe topographic constraints by its location at the foothills of the Alborz Mountains. Tehran is a very dynamic city which grew rapidly over the last decades. Being an Iranian it was therefore very logical for Dr. Jamal Jokar Arsanjani to choose the capital of his home country as a study area and at the same time a city that has to cope with all the problems of urban sprawl.

The original focus of Dr. Jokar Arsanjani's work is on agent-based modeling to predict land cover change for the Tehran area. This alone would already have been an interesting endeavor worth investigating. However, a real value of the work lies also in the extensive application and comparison of traditional methods to predict land cover change. These methods are cellular automata, Markov chain model, cellular automata Markov model, and the hybrid logistic regression model. In his thesis all these methods have been applied to the Tehran area to analyze and predict land cover change. In this respect the work can also serve as a text explaining the different approaches in their theoretical characteristics and practical applications. It is a particular value that the advantages and disadvantages of these methods are clearly exposed and explained.

Based on the preliminary findings of the different methods, finally, an agent-based model was developed that consist of government agents, developer agents, and resident agents, in order to simulate land cover change. Various parameters and behaviors were modeled and programmed in the ArcGIS environment. Since almost nothing in the real world follows a crisp classification, many traditional approaches suffer from a lack of adequately representing the real world situation. Fuzzy logic is one way to introduce uncertainty and vagueness to spatial analysis. Dr. Jamal Jokar Arsanjani uses fuzzy membership functions for the relevant factors in his geo-simulation research to represent a more natural behavior of the agents. This offers a more realistic analysis and provides results that better suit a real world situation.

The major value of this work is twofold: it shows a detailed comparison of existing methods for land cover change modeling, and it presents a novel approach in geo-simulation by applying agent-based modeling in a fuzzy setting. The thesis has already spawned several journal papers and Dr. Jokar Arsanjani's approach opens new perspectives for scientific problems in environmental monitoring, modeling and change detection.

Vienna, June 2011 Prof. Dr. Wolfgang Kainz

Contents

Abbreviations

ABM	Agent-based modelling
ABMS	Agent-based modelling simulation
AHP	Analytic hierarchy process
AI	Artificial intelligence
CA	Cellular automata
CBD	Central business district
CR	Consistency ratio
ESRI	Environmental systems research institute
FGE	Fuzzy geographical entities
GAL	GenePix array list
GAS	Geographic automata systems
GDP	Gross domestic product
GIS	Geographical information systems
GUI	Graphical user interface
LUCC	Land use/cover change
LULCC	Land use land cover change
MAS	Multi-agent systems
MASON	Multi-agent simulation of neighbourhood
MCE	Multi criteria evaluation
OBEUS	Object-based environment for urban simulation
Repast	REcursive porous agent simulation toolkit
RepastJ	Repast for Java
Repast.NET	Repast for Microsoft.NET
RepastPy	Repast for Python
RepastS	Repast Simphony
ROC	Relative operating characteristic
RS	Remote sensing
WGS84	World Geodetic System 84
WWW	World Wide Web

Chapter 1
General Introduction

1.1 Introduction

Land use/cover change is a complex matter, which is caused by numerous biophysical, socio-economical and economic factors. An obvious form of land use change in the suburbs of the metropolis is defined as urban sprawl. There are a number of techniques to model this issue in order to investigate this topic. These models have been developed since the 1960s and are increasing in terms of quantity and popularity. Some of these models suffer from a lack of consideration of some significant variables. The traditional methods (e.g. Cellular Automata, the Markov Chain Model, CA-Markov Model, and Logistic Regression Model) have some inherent weaknesses in consideration of human activity in the environment. The particular significance of this problem is the fact that humans are the main actors in the transformation of the environment, and impact upon the suburbs due to their settlement preferences and lifestyle choices. The main aim of this thesis is to examine some of those traditional techniques in order to discover their considerable advantages and disadvantages. These models are compared against each other to evaluate their functionality.

Benenson and Torrens (2004) the authors of the "Geosimulation: automata-based modelling of urban phenomena" believe and propose an innovative approach towards natural phenomena modelling, which they suggest is vastly turning to geospatial-explicit studies in the field of Geographic Automata System (GAS) modelling. In this particular research, the main goal is to introduce a new modelling system as an innovative paradigm in urban complexity by a GIS integrated automata system, the so-called geosimulation method as put forward by Benenson and Torrens (2004). This concept of geosimulation is based on geographically-related automata.

Updated and precise GIS and remote sensing databases serve as the primary information source for geosimulation implementation. Computational implementation of such geosimulation models is basically performed through object-oriented

J. Jokar Arsanjani, *Dynamic Land-Use/Cover Change Simulation: Geosimulation and Multi Agent-Based Modelling*, Springer Theses,
DOI: 10.1007/978-3-642-23705-8_1, © Springer-Verlag Berlin Heidelberg 2012

programming. Also, modern system theories provide the paradigmatic basis and analytical tools for investigating geosimulation models. In recent years, because of the rapid economic growth of developing countries, research in the phenomenon of urban expansion has increased exponentially. In contrast to regional models of 1980s, the 'new wave' of high-resolution models focuses on behaviour and transformations of urban objects (Hatna and Benenson 2007). Historically cities are complex systems and frequently evolve over time. Each singular activity and behaviour of the elements of this evolutionary system influences the decisions made by internal and external forces. Thus, each agent that might affect this system has, perforce, to be investigated for the simulation process (Crooks 2006). In addition, land use and land cover change modelling is an important and fast growing scientific field—because land use change is one of the most significant ways humans influence the surrounding environment. This issue is so extremely important that scientists have formed an international organisation known as "LUCC". The main thrust of this organization is its concern with the International Human Dimensions of Global Change Program and the International Geosphere Biosphere Program (Ellis and Pontius 2006; Lambin and Geist 2006; Pontius and Chen 2006).

Three main aims will be followed by this research: firstly, to create, modify and perform an agent-based modelling approach upon land use and cover change matter to evaluate the performance of this technique. More importantly this technique has not been imported into the GIS environment for simulation purposes. Therefore, the priority of this research is to construct an agent-based model in the interior of GIS software to present it as a new reliable system for GIS users. This method is being carried out for several purposes, such as traffic modelling (Ljubovic 2009), fire propagation (Michopoulos et al. 2004), complex behaviour modelling, urban growth and pedestrian movement (Kerridge et al. 2001).

Secondly, the land use and cover change subject was chosen for this agent-based modelling implementation because of the following motivations:

- The comparing of traditional methods with this proposed method in land use and land cover changes studies;
- The gathering of the results of each particular model to state an overall conclusion;
- The evaluation of advantages and disadvantages of each particular model for resultant improvement or hybrid model creation.

Therefore, the preliminary outcomes will be able to empower agent-based modelling as an approach to deduce benefits from each model's strength.

Thirdly, the constructed agent-based model will be able to simulate any forthcoming changes within a particular time period.

1.2 Problem Statement

1.2.1 Rapid Urban Expansion of Tehran

In developing countries, the population growth is principally rapid in the urban areas. Rapid urbanisation is consuming the farming land by urban built-up areas. Additionally, metropolitan population outside cities has increased faster than downtown areas in many regions, indicating a significant tendency of the outward extension of urban areas. Indeed, many cities are quickly growing at their fringes, swallowing rural areas and farming lands and converting into dense commercial and industrial areas (Huang et al. 2009).

The metropolis of Tehran, with around 13 million inhabitants (Iranian National Statistics Center 2006) is surrounded by Alborz Mountains in the north and Dasht-e Kavir in the south. It is located on a vast mountain slope with an altitude of 900–1,700 m above sea level. There are many cities remarkably close to Tehran which form the metropolitan area; the largest one is Karaj city, with more than one million inhabitants, 40 km away to the west, and the second largest city is Islamshahr with a population exceeding 300 thousand located 60 km to the south. These two cities also have their own suburb area. Moreover, there are several small towns and villages in the vicinity of Tehran in the situation of turning into large cities and then joining the metropolitan area. Tehran is limited in northern and eastern parts by high mountains that interrupt the urban expansion in these two directions.

Tehran has a rapid expansion rate and its sharp population growth in the recent decades has had many unpleasant impacts on the environment. From 1980 to 2000, resident population in Tehran nearly doubled. The physical growth of the city is replacing other land cover classes such as farming and open lands. Nearly 98.7% of the population of the metropolitan area lived in Tehran city 20 years ago, but within the recent years, it has decreased down to 67%. Moreover, about 33% of the population has moved to the suburbs, because of difficulties such as land prices and traffic and transportation problems. This process is changing urban areas that there is no significant boundary between urban and suburb areas. This challenges the urban planners and managers with new affairs on the administrative level. This growth in the metropolis is expanding and can result in more unsolvable complexities as other mega cities have faced before (e.g. Mexico City).

The Tehran growth has been becoming a national disaster, therefore massive immigration towards the city has to be stopped. Furthermore, this matter has caused remarkable damages in terms of environmental and economic aspects. As a matter of fact, Tehran province is the centre of accessibility to northern recreational facilities and its vast population is capable of damaging that area as well as increasing the speed of change in surroundings. Besides, establishment of Karaj province in 2010 in the vicinity of Tehran only 35 km away has also its own consequences that influence the growth rate. Consequently, the vast environmental damage of this decision cannot be ignored.

1.2.2 Limitations of Previous Approaches

It is essential for urban planners and land policy makers to focus on the trend of urban sprawl in the fringe of Tehran and its environmental impacts through the most reliable technique. Such a simulation will allow them to know about the probable future changes. Therefore, the direction and quantity of changes will become clear. So far, several methods about land change modelling have been performed in the Tehran metropolitan area by means of original and hybrid Cellular Automata Models, the Markov Chain Model and other artificial intelligence integration.

In recent years, inventive artificial intelligence prototypes for instance, geosimulation, agent-based modelling in contribution of fuzzy logic research have reached the capability to improve the quality and accuracy of such models (Rana and Sharma 2006). Land change researchers have been carrying out different methods and each one has some strengths and weaknesses which influence their results. Therefore, it is complex to compare the performance of the various models because the LUCC models have different fundamental structures. For instance, some models, such as the Cellular Automata, simulate changes in a binary form (i.e. between two land categories), whilst other models such as the CA-Markov, can simulate change among several categories (Pontius and Chen 2006).

On the other hand, some models are static (i.e. non dynamic) and some others have the capability of producing change probability surfaces for the allocation process at any time. In addition, a comprehensive comparison between different models in a particular study area has not been reported. This thesis aims to implement some models in a particular study area and conclude the advantages of each particular model. Also, in recent years, there are some software for implementing these approaches in both raster-based and vector-based data, but there is no valid literature to evaluate their quality and proficiency in the simulation process. Thus, we will draw a conclusion about them as well.

1.3 Research Hypotheses

In order to simulate the land use and cover changes by the geosimulation scenario and to compare this approach with traditional methods, the hypotheses of this research can be identified as follows:

- Geosimulation is a more applicable technique in comparison with other common techniques for land use change studies and prediction such as CA, Markov Chain and it is practical to replace it with other methodologies due to its individual characteristics in parameters modelling.
- Using different aspects of artificial intelligent approaches such as fuzzy logic, agent-based modelling and neuro-fuzzy systems in designing this simulation process and also in the prediction of future changes will be innovative.

1.4 Research Questions

As noted in Sect. 1.3, we intend to design various scenarios by means of traditional techniques and discover the advantages of each model and their strength to be utilised for designing agent-based model. Moreover, the land use change assessment process needs to evaluate the happened and probable changes in two different types of measurement; the quantity of change and the location of change. Therefore, these two values need to be assessed. Thus, the following research questions were designed for this study:

- *What are the potential limitations of common techniques for LUCC modelling? Are the MAS/LUCC models able to solve some of these constraints?*
- *What are the distinctive strengths of MAS/LUCC modelling techniques? How can these strengths conduct model developers in selecting the most appropriate modelling technique for their particular research question?*
- *Are MAS/LUCC model outcomes reliable in geospatially explicit studies?*
- *Do the agent-based models have the possibility to spatialize each particular variable in real-world phenomena?*
- *How can the ABM models be empirically parameterised, verified, and validated?*
- *Which type of agent is going to dominate the land change process in the study area?*

1.5 Research Objectives

In order to respond to the aforementioned research questions in Sect. 1.4, multiple scenarios for land use change modelling have to be designed. These scenarios comprise implementation of the Cellular Automata Model, the Markov Chain Model, the Cellular Automata-Markov Model and the Logistic Regression Model. Therefore, the outcomes of these models can lead this research to discover the appropriate drivers of change in the study area. The drivers of change can result in defining different agents and specifying their proper behaviours. These defined behaviours control each agent particularly and also the external interaction between all agents.

The main aims of this research in detail are listed below:

- To propose a generic method that can be followed to develop multi-agent systems in the GIS environments in various types of natural phenomena modelling,
- To design an agent-based modelling prototype based on geographic data and GIS functions, as well as to promote the capability of GIS environments' functionality for this matter,

- To propose an analysis technique to examine the results arising from the geo-simulation performance in comparison with other methodologies such as CA, Markov Chains and hybrid models,
- To consider the possibility of integrating GIS functions with ABM functions in GIS environment and segregate geosimulation from the ABM environments,
- To predict the future changes within a particular period through a customised scenario.

1.6 Research Approach

In order to achieve the noted objectives in Sect. 1.5, it was intended to discover the advantages and disadvantages of each existing model and therefore, feed the strengths of each model to the final ABM scenario. This thesis proposes an approach to create spatially explicit agent-based models by means of creating several relevant agents separately to simulate each one's behaviours independently. These agents are taught how to interact with other agents and themselves. Thus, the appropriate agents responsible for land change will be described by significant variables associated with each agent. Therefore, the following datasets were utilised as research materials:

- Satellite images such as Landsat data products from 1986, 1996 and 2006,
- Temporal land use/land cover maps,
- A comprehensive geodatabase of all geospatial variables in the study area (e.g. urban transportation data, land quality, building block details, demography statistics, land price data and other relevant data which will be explained in Chap. 4).

In addition, the research approach comprises ten main steps explained in more detail in the following chapters:

- *Multi-temporal land use mapping*
- *Implementation of the traditional approaches*
- *Designing a geosimulation model*
- *Comparison and evaluation of approaches*
- *Evaluation of current toolkits and software*
- *Execution of the designed geosimulation model*
- *Model evaluation*
- *Scenario customisation*
- *Analysis of outcomes from model implementation*
- *Prediction of future land use change.*

1.7 Organisation of the Thesis

This thesis consists of the following eight chapters as are listed below.

Chapter 1; *General Introduction* that presents a brief overview of the outlines of this research such as research hypotheses, research questions, research objectives, and the proposed approach.

Chapter 2; *Literature Review* that contains the scientific review of previous research carried out in the field of multi agent-based modelling approaches. Also, the role of artificial intelligence, computer modelling agents and GIS knowledge-based strategies in land use change studies will be discussed.

Chapter 3; *Study Area Description* brings a detailed description of the study area. This detailed information comprises a geographical explanation as well as a socio–economic description. Also, the importance of exploring land use change trends in the study area will be discussed.

Chapter 4; *Data Preparation* provides a comprehensive description about available data, required toolkits and software to run an agent-based model. The efficiency of several toolkits for this purpose will be evaluated in this chapter. An appropriate platform will be chosen which has enough capacity to satisfy our expectations for designing the ABM.

Chapter 5; *Implementation of Traditional Techniques* presents the traditional methodologies that have been employed in the field of land use change modelling (Cellular Automata, Markov Chain Model, CA-Markov Model, and Logistic Regression). These models will be designed to obtain their outputs in order to validate them as well as their results. The reasonable results will be taken into account in order to integrate their scientific background in our ABM.

Chapter 6; *Designing and Implementing Multi Agent Geosimulation* presents how the multi-agent simulation was developed. This chapter contains the followed steps to develop the ABM. The methodology of specifying the predefined agents with their preferences to settle will be explained.

Chapter 7; *Analysis of Results* presents how much the appropriate methodology is successful in achieving satisfactory results. In this chapter, a comparison between possible approaches and proposed ABM method will be presented. Additionally, a detailed and comprehensive discussion dealing with different scenarios considering their results will be presented. Uncertainty of utilised data and models will be noted.

Chapter 8; *Conclusions and Recommendations* illustrates an overall conclusion about the strengths and weaknesses of the implemented models. The original guidelines arising from this investigation will be depicted as well. This chapter will conclude the probable future works based on achieved outcomes.

References

Benenson I, Torrens PM (2004) Geosimulation: automata-based modeling of urban phenomena. Wiley, New York

Crooks AT (2006) Exploring cities using agent-based models and GIS. In Proceedings of the agent conference on social agents: results and prospects, University of Chicago and Argonne National Laboratory, Chicago, 2006

Ellis E, Pontius RG Jr (2006) Land-use and land-cover change—encyclopedia of earth. http://www.eoearth.org/article/land-use_and_land-cover_change

Hatna E, Benenson I (2007) Building a city in vitro: the experiment and the simulation model. Environ Planning B: Planning Des 34(4):687–707

Huang B, Zhang L, Wu B (2009) Spatiotemporal analysis of rural-urban land conversion. Int J Geog Inf Sci 23(3):379–398

Iranian National Statistics Center (2006) http://www.amar.org.ir

Kerridge J, Hine J, Wigan M (2001) Agent-based modelling of pedestrian movements: the questions that need to be asked and answered. Environ Planning B 28(3):327–342

Lambin EF, Geist HJ (2006) Land-use and land-cover change: local processes and global impacts. Springer, Berlin

Ljubovic V (2009) Traffic simulation using agent-based models. In information, communication and automation technologies, 2009. ICAT 2009. 22nd international symposium on information, communication and automation technologies, pp 1–6, 2009

Michopoulos J, Farhat C, Houstis E, Tsompanopoulou P, Zhang H, Gullaud T (2004) Agent-based simulation of data-driven fire propagation dynamics. In: Michopoulos J (ed) Agent-based simulation of data-driven fire propagation dynamics. Computational Science-ICCS 2004, pp 732–739

Pontius RG Jr, Chen H (2006) GEOMOD modeling, IDRISI Andes help contents. Massachusetts Clark University, Worcester, MA

Rana S, Sharma J (2006) Frontiers of geographic information technology, 1st edn. Springer, Berlin

Chapter 2
Literature Review

2.1 Introduction

In this chapter, it is intended to bring a summary about theoretical and fundamental fraction of agent-based modelling and how to design it according to the standard definitions. After this overview, the relationship between land change matter and the change drivers will be identified in terms of environmental and socio-economically investigation. Therefore, the appropriate and the most useful tools to implement the aim of this research will be depicted. It begins with the definition of the terms "land use" and "land cover" to outline their differences (Lambin et al. 2007). Land use/cover changes have various causes and consequences (i.e. loss of biodiversity, climate change, pollution, etc.) in the life cycle, which will be addressed briefly.

2.2 Land Use/Cover Change

The terms *Land use* and *Land cover* are not technically synonymous; hence, we draw attention to their unique characteristics to differentiate between them. The terms *land use* and *land cover* will be clarified in this chapter. There are different definitions of land cover and land use among the relevant scientists. Therefore, a brief explanation about these two terms is provided in this section from the Encyclopaedia of Earth. In general, the term land use and land cover change (LULCC) identifies all kinds of human modification of the Earth's surface. *Land cover refers to the physical and biological cover over the surface of land, including water, vegetation, bare soil, and/or artificial structures* (Ellis and Pontius 2006).

Land use has a complicated expression with different views compared with the term land cover. In fact, social scientists and land managers characterise this term

J. Jokar Arsanjani, *Dynamic Land-Use/Cover Change Simulation: Geosimulation and Multi Agent-Based Modelling*, Springer Theses,
DOI: 10.1007/978-3-642-23705-8_2,

more general to involve the social and economic purposes. Natural science researchers classify the term land use in different aspects of human activities upon lands such as farming, forestry and man-made constructions.

TurnerII et al. (1995) believe *Land use involves both the manner in which the biophysical attributes of the land are manipulated and the intent underlying that manipulation—the purpose for which the land is used.* Lambin et al. (2007) differentiate between *land cover* (i.e. whatever can be observed such as grass, building) and *land use* (i.e. the actual use of land types such as grassland for livestock grazing, residential area). In fact, the term *land use/cover* will be used chiefly in this thesis, referring to the land cover and the actual land use.

2.3 Land Use/Cover Change Causes and Consequences

LUCC can occur through the direct and indirect consequences of human activities to secure essential resources. This may first have occurred by means of burning of areas to develop the availability of wild game and it accelerated with the birth of agriculture, resulting in extensive clearing such as deforestation and earth's terrestrial surface management that takes place today (Ellis and Pontius 2006). Land-use/cover change is known as a complex process which is caused by the mutual interactions between environmental and social factors at different spatial and temporal scales (Valbuena et al. 2008; Rindfuss et al. 2004).

More recently, industrial activities and developments, the so-called industrialisation, has encouraged the concentration of population within urban areas. This is called urbanization, which includes depopulation of rural regions along with intensive farming in the most productive lands and the abandonment of marginal lands (Ellis and Pontius 2006). Land use changes are increasingly known as the consequence of actors and factors' interactions (Bakker and van Doorn 2009). These conversions and their consequences are obvious around the world and it has been becoming a disaster around the metropolitan areas in developing countries.

2.3.1 Loss of Biodiversity

Biodiversity has been diminishing considerably by land change. While lands change from a primary forested land to a farming type, the loss of forest species within deforested areas is immediate and huge (Ellis and Pontius 2006). According to Ellis and Pontius (2006):

> The habitat suitability of forests and other ecosystems surrounding those under intensive use are also impacted by the fragmenting of existing habitat into smaller pieces, which exposes forest edges to external influences and decreases core habitat area.

2.3.2 Climate Change

Land use and cover change matters play a significant role in climate change at different scales such as regional, local and global scales. At global scale, LUCC is accountable for releasing greenhouse gases to the atmosphere, thus leading to global warming. LUCC is able to increase the carbon dioxide balance to the atmosphere by disturbance of terrestrial soils and vegetation. Furthermore, LUCC undoubtedly plays an essential role in greenhouse gas emissions.

2.3.3 Pollution

Tree harvesting, land clearing and other forms of biomass damage to the environment arising from land change are able to increase the pollution percentage of the environment. Vegetation removal makes soils vulnerable to a massive increase in windy and water soil erosion forms, particularly on steep topography. When accompanied by fire, also pollutants to the atmosphere are released. Soil fertility degradation within time is not the only negative impact; it does not only cause damage to the land suitability for future farming, but also releases a huge amount of phosphorus, nitrogen, and sediments to aquatic ecosystems, causing multiple harmful impacts. All of these issues drive water, soil and air pollution at large scale. Besides, other agricultural activities such as using herbicides and pesticides also release toxics to the surface waters, which sometimes remain in the top soil.

2.3.4 Other Impacts

Other environmental impacts of LUCC include the destruction of stratospheric ozone by oxide release from agricultural land and altered regional and local hydrology. Moreover, the most urgent concern for a great part of the human population and most governments is the long-term supply and production of food and other fundamentals required in the future Pontius and Chen (2006).

2.4 Driving Forces of the Land Use/Cover Changes

Assessing the driving forces behind LUCC is essential if previous patterns can explain and be utilised in forecasting future patterns. Land use and cover change can be caused by multiple driving forces that control some environmental, social and economic variables. These driving forces can contain any factor which influences human activities, including local culture, economic and financial matters, environmental circumstances (i.e. greenness, land quality, terrain situation, water availability and accessibility to recreation), current land policy and development plans, and also interactions between these factors. Therefore, these drivers have to be found to pursue these controlling variables. The driving forces will be utilised in order to manage land change.

Investigation of interrelations between the drivers of land change needs a strong knowledge about methods and effective variables, as well as land policy (Ellis and Pontius 2006). LUCC is frequently addressed through various selected biophysical and socioeconomic variables. In order to facilitate simulation, driving factors are mostly considered exogenous to the land use system (Verburg et al. 2004). Associations between driving forces and LUCC could be addressed qualitatively and quantitatively by means of appropriate approaches.

2.5 Land Use/Cover Change Simulation

Spatially-explicit models, which consider social and environmental causes and consequences, can be the most appropriate form of existing models to simulate land changes. These approaches are capable of checking relationships between environmental and social variables. Integration of existing geographical data and advanced GIS functionality, as well as the ABM functionalities allow this research to achieve the proposed objectives. Considering this, LUCC can be affected remarkably by political and economic decisions. However, the traditional models are not capable of considering all these variables (Ellis and Pontius 2006). These geospatial models can result in precise outcomes that help land managers and policymakers towards a better landscape administration and sustainable land management.

It does not seem simple to compare the performance of the numerous models of LUCC modelling, because they are created from different fundamental bases. For instance, the GEOMOD model simulates change between two land categories, whilst others, such as the Markov chain model and the cellular automata-Markov model simulate change among several categories. Nonetheless, by developing multiagent-based systems (MABS) lately, research is improving these methods to achieve better outcomes. Also, some models use raster data, while others are in vector format. Even in the case of all researchers using the same model, comparison among model performance would still be complicated because researchers usually focus on one study area and do not make a global use approach (Pontius and Chen 2006).

Pontius and Chen (2006) believe that,

> it is complicated to separate the quality of the model from the complexity of the landscape and the data.

As an example, if a model does not perform strongly, it does not necessarily imply that the conceptual foundation of that model is weak, but it could mean that the event of land change in that particular study area is complex or the data is inaccurate. However, if a model performs properly, it is difficult to recognize whether theoretical basis of that model is strong, or that land change case in the study area is particularly uncomplicated, or the used data is extremely uncertain.

Perhaps most importantly, there is not yet a global agreement about methods to determine the performance of LUCC models; therefore, two users who performed the same model on the same landscape and data situation might evaluate one simulation execution differently depending on the criteria used for evaluation (Pontius Jr. and Chen 2006). Land-use change modellers might conclude that the intellectual basis of the validation of the models has some weaknesses (Kok and Veldkamp 2001; Pontius Jr. et al. 2001; Pontius and Schneider 2001; Pontius et al. 2004).

2.6 Land Use Change Trend

Change in economy and spatial distribution of population can occur through conversion from one land use to another, for instance, converting farming lands into residential, industrial, commercial or recreational use. The land owners play a key role in whatever will take place at the land and, therefore, their decisions identify the direction and quantity of changes (Ettema et al. 2007).

Therefore, different types of land owners (e.g. farmers, developers, private individuals, government) decide in a different way according to their type and their parameters. The owners have to supply the financial investment of land change, thus, their awareness of the economic situation can control the speed of the changes. At each time step, the landowner can decide the following decisions:

- Leave the land at current circumstances;
- Develop the land by changing the land usage and exploit it;
- Develop the land by changing the land usage and sell it;
- Sell the land to another owner.

However, the options vary for some owners. For instance, a farmer is not able to develop his land into a residential area, if he does not have the required investment power and skills. Moreover, all actions may not be allowed given planning regulations. Ettema et al. (2007) differentiate between three different types of owners with preferences: farmers (preferences: exploit, sell or buy), government (preferences: maintain, sell to farmer, sell to developer or develop and maintain) and developers (preferences: develop and sell, redevelop and exploit, sell).

Eventually, the decision, which will be most likely made, totally depends on the expected value of each option to the owner. In case of commercial owners, utility will match with profitability: the action will be taken that delivers the highest profit. In case of governmental part, also social benefits might play a significant role, whereas in the farmers' case, personal and emotional reasons may influence their decision. The market price is a valuable index in deciding whether or not to sell a land with or without developing it (Ettema et al. 2007; Koomen et al. 2007).

2.7 Predicting Future Land Use Patterns

As an essential part of their profession, land use planners envision and forecast alternate future land use and activity patterns in order to change the status quo (Brail and Klosterman 2001). Assessing, forecasting, and evaluating future land change is a complex set of tasks and, hence, it has to be performed after a deep scientific knowledge of the extent individuals, characters, as well as consequences of land transformation have been gathered (Meyer and Turner 1994). A typical land use planning process requires the landscape planners to realise, classify, and investigate the current circumstances in order to project future probable development patterns, and propose plans based on available information (Brail and Klosterman 2001). According to Brail and Klosterman (2001), planners usually approach this task in two ways, a predominant or traditional approach and an analytical approach. The traditional approach foresees a future land use outcome and then prioritises present-day policies required to achieve that outcome. The analytical approach simulates alternate current strategies and compares their consequences.

A recent pervasive approach to consider and simulate human decisions in LUCC is the use of multi-agent systems (MAS) (Parker et al. 2003; Matthews 2006; Robinson et al. 2007; Valbuena et al. 2008). MAS are defined as modelling tools that allow entities to make decisions according to the predefined agents, and the environment also has a spatial explicit pattern. In fact, agents in the system might represent groups of people or individuals, etc. (Valbuena et al. 2008; Sawyer 2003; Bonabeau 2002; Crawford et al. 2005). Agents can be designed with different characteristics which will be explained later in this chapter.

2.8 Simulating Sprawl

Urban sprawl is fairly a contemporary theme in urban studies. Torrens (2006a) noted that;

> Suburban sprawl is among the most important urban policy matters facing contemporary cities.

Spatial simulation is able to support sprawl associated research by means of what-if experimentation environments. Sprawling cities are being considered as complex systems and this justifies use of geosimulation to accommodate the space–time dynamics of numerous interacting entities. Automata are compatible tools to represent such systems, but they can be improved to capture uniquely geographical traits of phenomena such as sprawl. Therefore, the development of a model for the geographic dynamics simulation of urban sprawl is explored (Torrens 2006a).

2.9 Approaches to the LUCC Modelling

There are plenty of models concerning land use/cover change modelling. Despite their differences they basically rely on a limited number of methods and assumptions. Those models include economic models (Irwin and Geoghegan 2001), spatial interaction, cellular automata (Yang et al. 2008), statistical models (Veldkamp and Lambin 2001), optimisation techniques (e.g. Ducheyne 2003), rule based models, multi-agent models (e.g. Torrens 2006b), and microsimulation (e.g. Timmermans 2003).

This subsection aims to bring an overview of traditional and current LUCC modelling techniques and eventually, will suggest multi-agent-based systems as a complementary tool. Briefly, the strengths and weaknesses of some models will be discussed here. This appraisal is not in-depth and only presents the best methods which can be complemented by MAS models:

- Equation-Based Models,
- System Models,
- Statistical Techniques,
- Expert Models,
- Evolutionary Models,
- Economic Principles,
- Spatial Interaction,
- Evolutionary Algorithms,
- Genetic Algorithms,
- Optimisation Techniques,
- Cellular Models,
- Hybrid Models,
- Multi-Agent Models,
- Microsimulation.

2.10 Agent-Based Modelling and Geosimulation Terminology

Macal and North (2006) believe that “There is no universal agreement on the precise definition of the term ‘agent’, although definitions tend to agree on more points than they disagree”. It seems very complicated to extract agent characteristics from the literature in a consistent and constant perspective, because they are utilised in different ways (Bonabeau 2002).

Agent-based modelling (ABM) is able to simulate the individual activities by measuring their behaviour and results over time for developing models of cities (Crooks 2006). Crooks (2006) explains cities as follows:

> Cities are complex systems, with many dynamically changing parameters and large numbers of discrete actors. The heterogeneous nature of cities, make it difficult to

> generalize localized problems from that of city-wide problems. To understand cities' problems such as sprawl, congestion and segregation, we need to adapt a bottom-up approach to urban systems, to research the reasoning on which individual decisions are made. As cities are highly dynamic, both in space and time and secondly, as cities operate on a cross scale basis, propagating through urban systems from interactions between individuals in space to regional scale geographies. For example, it is easier to conceptualize, and model how individual vehicles move around on a road network, where each car follows a simple set of rules. For instance if there's a car close ahead, it slows down, if there's no car ahead, it speeds up and how this can lead to traffic jams without any obvious incident.

Human agents are becoming increasingly significant in land use simulation, despite the fact that traditional environmental and economic models presume one main agent aiming at optimisation in financial conditions (Bakker and van Doorn 2009; Irwin and Bockstael 2002). A variety of MAS models has been developed for land use dynamics modelling that will be mentioned in this chapter (and so far, these models have mostly been performed rule-based (Ligtenberg et al. 2004; Bousquet and Le Page 2004; Berger 2001; Bakker and van Doorn 2009). Certainly, it is vital to represent the agents' intentions and behaviours with respect to decision making, realistically.

2.10.1 Agents and Agent-Based Models

An agent can be defined according to Russell and Norvig (2009) as follows:

> The concept of an agent is meant to be a tool for system analyzing, not an absolute classification where entities can be defined as agents or non-agents.

For instance, a number of experts take into consideration any sort of independent components (e.g. software, individual, etc.) an agent, while some others believe that a component's behaviour needs to be adaptive in order to be considered an agent, where the term agent is reserved for components that can learn through their environments and change their behaviours accordingly (Macal and North 2005). Nevertheless, several common features exist for most agents (Wooldridge and Jennings 1995; Castle and Crooks 2006)—extended and explained further by Franklin and Graesser (1996), Epstein (2007), and Macal and North (2005).

Therefore, the following characteristics can be defined according to the definitions by Benenson and Torrens (2006).

- *Autonomy*: Agents are independent and autonomous units that are capable of information processes and exchanging them with other agents to independently make decisions. They are also capable of being interactive with other agents and this does not necessarily influence their autonomy (Castle and Crooks 2006; Smith et al. 2007; Benenson and Torrens 2004).

- *Heterogeneity*: Agents can exist and act as groups, but they are constructed through a bottom-up way and combinations of similar autonomous individuals.
- *Mobility*: The mobility of agents is particularly a practical characteristic for spatial simulations. Agents can move around the space within a model.
- *Adaptation and Learning*: Agents are flexible to be adaptive to produce Complex Adaptive Systems (Holland 1996). Agents can be designed to change their locations depending on their current state, following their designed memory (Smith et al. 2007).
- *Activity*: Agents have to be active since they perform independent impacts in a Geosimulation. The following active features can be identified:
  - *Pro-active* (*i.e. goal-directed*): Agents are often considered goal-directed elements, following goals to be accomplished with respect to their behaviours. For instance, agents in a geographic environment can be designed to discover a set of spatial manipulations to achieve an aim within a certain limitation (e.g. time), while evacuating a building during an urgent situation.
  - *Reactive* (*i.e. perceptive*): Agents can be developed to have a consciousness of their surroundings to draw a 'mental map' by means of prior knowledge; thus, providing them with an awareness of other entities, obstacles, or required destinations within their environment.
  - *Bounded Rationality*: In social sciences, a dominant type of modelling based on rational-choice paradigm has to exist. Rational-choice models commonly assume that agents are perfectly rational optimisers with easy access to gathered information, foresight, and infinite analytical capability. These agents are therefore able to solve deductively complex mathematical optimization matters.
  - *Interactive* (*i.e. communicative*): Agents communicate to each other, extensively. For instance, agents can enquire other agents and the environment within a neighbourhood, searching particular attributes, with the ability to disregard an input which does not match a desirable threshold.

Agent-based models consist of several interactive agents placed within a simulation environment. Relationships between the existing agents are formulated, linking agents to other agents within a system. Relationships can be specified in a number of ways, from simply reactive (i.e. agents only accomplish events when activated to do so by external stimulus e.g. behaviour of another agent), to goal-directed (i.e. seeking a particular purpose). In some cases, the action of predefined agents can be programmed to occur synchronously (i.e. each particular agent executes events at each discrete time point), or asynchronously (i.e. agent reactions are planned by the actions of other agents and/or with reference to a predefined time) (Showalter and Lu 2009).

According to Castle and Crooks (2006),

> Environments define the space in which agents operate, serving to support their interaction with the environment and other agents. Agents within an environment may be spatially explicit, meaning agents have a location in geometrical space, although the agent itself may be static. For example, within a building evacuation model agents would be required

> to have a specific location for them to assess their exit strategy. Conversely, agents within an environment may be spatially implicit; meaning their location within the environment is irrelevant. For instance, a model of a computer network does necessarily require each computer to know the physical location of other computers within the network.

In simulation environments, agent-based models can be used as experimental media for performing and monitoring agent-based simulations. They can be pictured as a miniature laboratory, where the characteristics and behaviours of agents can be transformed and the consequences observed over multiple simulation runs. As a matter of fact, the ability of individual actions simulation upon various agents and measure the resulting system behaviour and consequences over time means agent-based models can be employed profitably in order to investigate processes that operate at various scales. In fact, the roots of ABMs lie within the individuals' behaviours simulation and human decision-making (Bonabeau 2002).

Furthermore, Bazghandi and Pouyan (2008) state that

> ABM is not the same as object-oriented simulation, although the object-oriented paradigm provides a suitable medium for the development of agent-based models. Consequently, ABM systems are invariably object-oriented.

Considering that agent-based models express the behaviours and interactions of a system's constituent parts from bottom to top, they are the canonical approach for modelling emergent phenomena (Bonabeau 2002). Bonabeau (2002) has categorised a number of conditions that ABMs are practical for capturing emergent behaviour.

2.11 Characteristics of the Geosimulation Model

Geosimulation differs from cellular automata in one particular respect: individual automata are basically free to move around, i.e. they are not fixed agents and their movements do not have to take place cell by cell. This feature has obvious consequences for the representation of spatial systems (Longley and Batty 2003); therefore, this topic will be explained in detail within this chapter. Figure 2.1 represents a schematic view of characteristics of a multi-agent system.

2.11.1 Management of Spatial Entities

A basic aspect of geosimulation regards the characterisation of spatial entities that form the building blocks of a simulation model. In fact, urban simulation models have identified units of urban systems (e.g. real estate, land, individuals, etc.) by aggregation of geographic zones, tracts, and socioeconomic groups. These collect units that are spatially modifiable (Openshaw 1983).

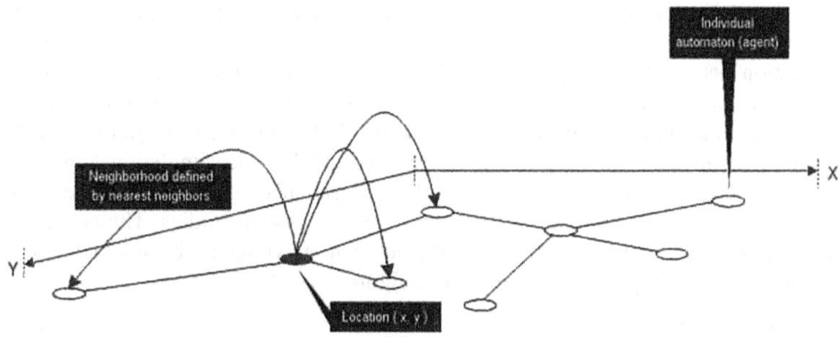

Fig. 2.1 A schematic view of a multi-agent system (Benenson and Torrens 2004)

2.11.2 Management of Spatial Relationships

The second aspect of geosimulation relates to the portrayal of spatial relationships in models. For instance, we can consider this in the framework of geospatial interactions; their representation in traditional urban simulation has been limited to flows between aggregate units. Geosimulation models consider interactions as an outcome of the behaviour of elementary geographic objects. In this way, geosimulation models have the potential to represent spatial interaction of a much wider spectrum of forms, including traditional and far-distance migration (Crooks 2006).

2.11.3 Management of Time

The third distinctive characteristic of them relates to the action of time in models. Urban systems convert over time, and diverse phenomena happen at different time scales. Benenson and Torrens (2004) believe that

> Geosimulation models treat time through intuitively justified units such as housing search cycles. Objects' temporal behaviour can be considered as either synchronous, when all objects change simultaneously, or asynchronous, when they change in turn, with each observing the urban reality as left by the previous one.

2.11.4 Direct Modelling

Disappointment with the appearance of *urban simulation* as a new field of study in the 1970s was an expectation of what urban simulation models need to accomplish in reality (Crooks 2007a, b; Batty 2005). One of the goals of the geosimulation approach is to move towards the creation of "tools to think with" (Benenson and Torrens 2004). Benenson and Torrens (2004) note that

Realistic descriptions of objects' behaviour in ways that were not previously obtainable, either technologically or intellectually, makes these worthwhile and, further, allows for direct relation between conceptual and real-world modelling. The idea underlying geosimulation is that the same model can be applied to abstract real-world phenomena; if modelled phenomena are an abstraction of real-world phenomena, why should modelled objects differ from their counterparts in the real world? The Geosimulation approach is supported by several key developments in the geographical sciences and other fields, particularly mathematics, computer science, and general system studies. The cornerstone of the geosimulation approach, however, is the automaton, which has been widely used in simulation and features prominently in geosimulation toolkits.

2.12 The Basic of Geosimulation Framework: Automata

The description of objects' behaviour in the geosimulation framework is based on the idea of automata. Simply stated, an automaton is a processing mechanism with characteristics that change over time based on its internal characteristics, rules, and external input. Automata are used to process information input into them from their environs with the characteristics altering according to rules that govern their reaction to those inputs. Levy (1992) explains automata as below.

> An automaton is a machine that processes information, proceeding logically, inexorably performing its next action after applying data received from outside itself in light of instructions programmed within itself.

Automata are a practical concept of "behaving objects" for many causes, but chiefly because they provide an efficient formal mechanism for representing their fundamental properties: behaviours, attributes, relationships, environments, and time.

Formally, a finite automaton A can be represented by means of a finite set of states $S = \{S_1, S_2, S_3, \ldots, S_N\}$ and a set of transition rules T.

$$A \sim (S, T) \tag{2.1}$$

Transition rules define an automaton's state, S_{t+1}, at time step $t + 1$ depending on its state, $S_t(S_t, S_{t+1} \in \{S\})$, and input, I_t, at time step t:

$$T : (S_t, I_t) \rightarrow S_{t+1} \tag{2.2}$$

2.13 Cellular Automata versus Multi-Agent Systems

Geosimulation requires a geospatial structure for modelling urban systems, one as formulated on the basis of objects located in space. Ideally, such an approach allow for simulated geospatial entities to be considered as automata; moreover,

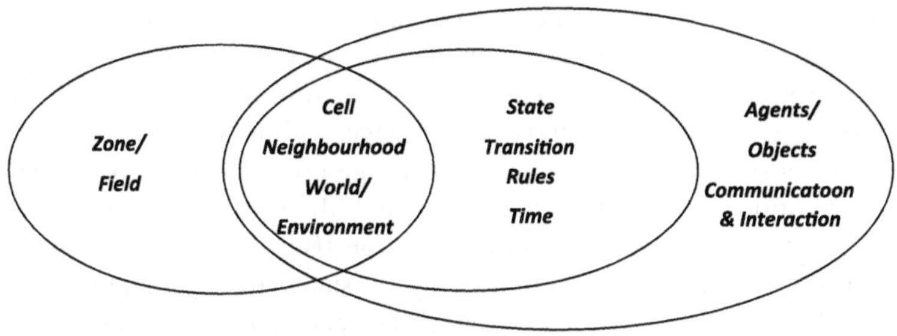

Fig. 2.2 Relation between cell-based GIS, CA modelling and MAS

Cellular Automata and multi-agent systems concepts could ideally be combined by considering collections of interacting geographic automata. In this section of the chapter, it is intended to introduce such a framework, which considers geographic objects, interacting to form geographic automata systems and urban system as a whole are considered as the products of collective dynamics among multiple inanimate and animate geographic automata (Benenson and Torrens 2004). Figure 2.2 represents relation between cell-based GIS, CA modelling and MAS.

2.14 Geographic Automata System

The geographic automata system (GAS) framework joins CA and MAS directly reflecting a geographic and object-based (more particularly, automata-based) view of urban systems. This idea was introduced for the first time by (Benenson and Torrens 2004) as a new paradigm in natural studies for better and more accurate results.

2.14.1 Definitions of Geographic Automata Systems

There is a distinct class of automata, geographic automata systems (GAS), consisting of geographic automata of various types. In general, the states and transition rules characterise automata (Benenson and Torrens 2004).

Basically, the G value in GAS can be defined as consisting of seven following components:

$$G \sim (K\ ;\ S,\ T_S;\ L,\ M_L;\ N,\ R_N) \tag{2.3}$$

Here, K represents a set of types of automata represented in the GAS and three pairs of symbols denote the other components, each one representing a specific spatial or non-spatial characteristic and the rules identify its dynamics. The first pair denotes a set of states S, linked with the GAS. G consists of a set of states S_k of automata for each type of $k \in K$. A set of state transition rules T_S, determine how automata positions are supposed to change within time. The second pair represents location information. L denotes the geo-referencing conventions that dictate the location of automata in the system and M_L denotes the movement rules for automata, governing changes in their location (Benenson and Torrens 2003). Hence, changes in location and transitions of states for geographic automata depend on the automata and also, on input (I), specified by the states of neighbours. The third pair specifies this condition. N denotes the neighbours of the automata and R_N represents the neighbourhood rules that manage how automata relate to the other automata in their vicinity.

2.14.2 Geographic Automata Types

GAS consists of different types of automata. Two main types of automata can be distinguished; non-fixed and fixed geographic automata. Fixed geographic automata stand for objects that do not move over time and thus have close analogies with CA cells. For instance, in the context of urban systems, a variety of urban items may be indicated as fixed geographic automata: building footprints, road networks conjunctions, parks, etc. Fixed geographic automata may be addressed by any of the transition rules outlined already, except rules of movement, M_L. Non-fixed geographic automata identify entities which move around over time. The full array of rules for GAS can perform with non-fixed geographic automata, including movement rules (Benenson and Torrens 2004).

2.14.3 Geographic Automata States and State Transition Rules

A number of state variables S can be assigned to the individual geographic automata, that comprise a GAS, and these states explain the characteristics of the automata. Any variable can be employed to derive state values, including variables of geographic significance. Pointing to the non-fixed automata, location variables of relevance to the transition rules of the model might be initiated.

In fact, state transition rules are based on geographic automata of all forms of K. It seems necessarily vital mentioning that, in the framework of the GAS, CA is artificially closed, simply because cell state transition rules are driven only by cells (Benenson and Torrens 2004). In contrast, the states of urban infrastructure objects represented by means of geographic automata totally depend on the surrounding objects of that infrastructure, but are also driven by mobile geospatial automata (i.e. agents) that are responsible for controlling object states such as land value or land-use (O'Sullivan et al. 2003). This is a crucial concept for simulating human-

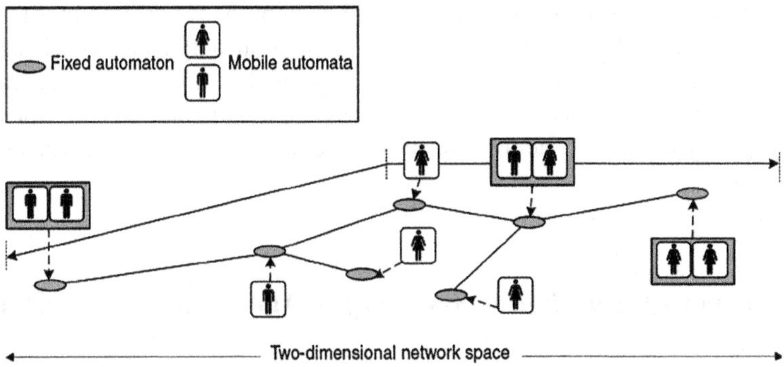

Fig. 2.3 Direct and indirect geo-referencing of fixed and non-fixed GA (Benenson and Torrens 2004)

driven urban systems that show how individuals interact and are affected by the environments.

2.14.4 Geographic Automata Spatial Migration Rules

Geo-referencing conventions (L) administrate how geographic automata should be registered in space. For fixed geographic automata, geo-referencing is a straightforward process in most instances; these automata can be geo-referenced by recording their position coordinates. However, for non-fixed geographic automata, geo-referencing has to be dynamic and automata may move. Also, their location in relation to other automata, represented in simulated goals, destinations, opportunities, etc., may be dynamic in space and time (see Fig. 2.3). It is also essential noting that there are examples in which Georeferencing is dynamic for fixed geographic automata also, for example, when land parcel objects are sub-divided during simulation (Benenson and Torrens 2004).

2.14.5 Geographic Automata Neighbours and Neighbourhood Rules

Another element of GAS that requires explicit explanation is the set of neighbours of automata, N, and the rule set for determining the change in neighbourhood relationships between automata, R_N. Different type of neighbours is necessary for the application of transition rules state transition (T_S) and movement (M_L), which totally depend on characteristics of geographic automata and their neighbours.

In opposition to the static and symmetrical neighbourhoods utilised in usual CA models, geographical relationships between geospatial automata can change in space and time, thus, RN rules need to be formulated to account for geographic automata positions' neighbours at each time point. Neighbourhood rules for fixed geographic objects can be defined easily comparatively, because the objects are static in space.

2.14.6 Types of Simulation Systems for Agent-Based Modelling

Generally, two types of simulation systems can be performed to develop agent-based models: either toolkits or software. Based on this, toolkits are simulation or modelling systems that provide a conceptual framework for designing ABMs which provide required libraries of software functionality that consist of pre-defined modules, routines and functions distinctively designed for ABM. The object-oriented prototype allows importing extra functionalities through other libraries, which are not supplied by the simulation toolkit, developing the capabilities of these toolkits (Crooks et al. 2008). The most interesting part of this approach is the capability of integration of GIS functionality from ArcGIS software libraries with an ABM context.

The development of agent-based models can be significantly facilitated by the utilisation of simulation and modelling toolkits. In fact, they are able to provide reliable templates for the design, accomplishment and visualisation of agent-based models, allowing modellers to concentrate on the content of research, rather than coding fundamentals required to run a simulation (Tobias and Hofmann 2004). In particular, the use of toolkits can decrease the burden of modellers challenged with programming matters of a simulation (e.g. GUI design, data import and export, visualisation and model representation). It is also crucial to improve the model's trustworthiness and efficiency (Smith et al. 2007).

Unsurprisingly, there are limitations of using simulation/modelling systems to develop agent-based models; for instance, a considerable amount of effort has to be spent to realise how to design and implement a model (Crooks 2007a, b).

Benenson et al. (2005) and Crooks (2007a, b) note that

> toolkit users are accompanied by the fear of discovering that a particular function cannot be used, will conflict, or is incompatible with another part of the model late in the development process.

2.15 Current Simulation Systems

Various environments are available in order to develop agent-based models. This section aims to review an overview of these systems:

1. Open source such as Swarm, MASON and Repast,
2. Shareware/freeware such as StarLogo, NetLogo and OBEUS,
3. Proprietary systems such as AgentSheets and AnyLogic (Bandini et al. 2009).

The mentioned systems need to fulfil the majority of the following criteria:

- retained and still being improved,
- broad range of users and also supported by strong user communities,
- accompanied by various demonstration models and in some instances the model's programming source code made available,
- Capable of developing spatially explicit models and integration with GIS functionality.

Further information about each system, as well as identifying examples of geo-spatial models that have been developed will be provided in this section. In this part of chapter, a brief introduction of all affordable toolkits will be presented in order to acquire a preliminary knowledge over mentioned prototypes. Certainly, the earliest and most well-known toolkit was SWARM, although many other toolkits more recently have been released. There are a variety of toolkits available for ABM at this time. However, variation between toolkits needs to be considered. For instance, their purpose, level of development, and modelling capabilities can vary. A review of the most user-friendly toolkits will be presented throughout this chapter.

2.15.1 ASCAPE

ASCAPE (Agent-Landscape) is one of the earliest toolkits associated with ABMs which has been developed by the Centre on Social and Economic Dynamics (CSED), Brookings Institution. ASCAPE is a research toolkit to support agent-based modelling and simulation. In fact, high-level frameworks support complex model designs, while end-user tools prepare it for non-programmers to investigate various aspects of model dynamics. This toolkit is written completely in Java, and runs on Java-enabled platforms. Models developed by this means can be easily published to the web for use with common web browsers (Batty and Jiang 1999; Epstein and Axtell 1996).

2.15.2 StarLogo

According to the StarLogo official website (2008);

> StarLogo is a programmable modelling environment for exploring the workings of decentralized systems–systems that are organized without an organizer, coordinated without a coordinator. With StarLogo, you can model (and gain insights into) many real-life phenomena, such as bird flocks, traffic jams, ant colonies, and market economies.

StarLogo is a particular version of the Logo programming language. Also, it is practical to create drawings and animations by giving commands to graphics. It expands this idea by allowing users to control many graphic *turtles* in parallel. In addition, StarLogo makes the turtles' world computationally active; therefore, it is possible to create the turtles' environment by code. Turtles and patches can interact with one another. StarLogo is predominantly well-suited for artificial life investigations. In decentralised systems, orderly patterns can take place without centralised control. StarLogo has been developed to facilitate students, as well as researchers to extend new ways of understanding decentralised systems (Camazine et al. 2003).

2.15.3 NetLogo

NetLogo is a multi-agent programmable platform developed by the Centre for Connected Learning and Computer-Based modelling, Northwestern University, USA (Tisue and Wilensky 2004). NetLogo allows the users to access a large library of sample models and code examples that help users to start authoring models. NetLogo is being used by research labs and university lessons in social and natural sciences.

2.15.4 OBEUS

Object-Based Environment for Urban Simulation (OBEUS) is a software environment based on a GAS conceptual core. In fact, OBEUS has been established according to the basic components of GAS with respect to automata types. These are accomplished by means of three following root classes:

- Population that contains information regarding the population of objects of a given type k as a whole;
- Geo-Automata, acting as a container for geographic automata of a given type k;
- Geo-Relationship that facilitates specification of spatial relationships between geographic automata of the same or different types (Benenson and Torrens 2004).

2.15.5 AgentSheets

AgentSheets is another toolkit for construction of interactive graphical systems. It is a simulation system that allows modellers with partial coding skill to develop an agent-based model, because models can be developed through a GUI

(Repenning et al. 2000). Several demonstration models exist on the system web-site; for instance, Sustainopolis. The system lacks, however, functionality to dynamically chart simulation output, and agents are limited to movement within a 2-dimension lattice environment (Crooks 2007a, b).

2.15.6 AnyLogic

AnyLogic allows modifying a simulation model using several methods; system dynamics, agent based and discrete event (process-centric) modelling. Furthermore, it is also possible to combine different methods in one model. AnyLogic modelling language is an extension of UML-RT, a set of the best engineering practices have been verified successfully in the modelling of complex systems (Anylogic 2006).

2.15.7 SWARM

Swarm is one of the oldest agent-based modelling toolkits. Swarm has been originally written in Objective-C language, and then exported to Java (Getchell 2008). Nevertheless, the documentation and research papers on Swarm established many of the foundational concepts and ideas in ABM, and reading over these materials serves as an excellent introduction to the large and growing field of agent-based modelling.

2.15.8 MASON

MASON or Multi-Agent Simulator of Neighbourhoods/Networks is another simulation library in Java, designed to serve as the base class structure for custom Java simulations. It also includes a model library and suite of 2D and 3D visualisation tools, and is developed with an emphasis on speed and portability.

2.15.9 NetLogo

NetLogo is another ABM toolkit, which is not open source, and also designed for educational use, being based on a simple Logo-type language. It was initially developed in 1999 by Uri Wilensky, and it has been under continuous development thereafter, and has a large and broad user community.

2.15.10 *Repast*

Repast has been made based on Swarm, but executed in Java. Repast has several versions available; the current standard Repast version is 3. RepastPy is a simplified version of Repast, and introduced a friendly graphical user interface. Also it benefits from a subset of Python as its scripting language. RepastPy is faster and easier to employ than Repast, and is generally recommended as being a good version for creating prototype models. In fact, the Python scripts generate Java objects (Getchell 2008).

Repast.NET is another version of Repast based on the .NET runtime. The .NET runtime is flexible and powerful due to having a large number of functional libraries for handling nearly anything. It also has a stylish successor language to Java, C#, as well as the ability to run any language that can be linked to the .NET platform such as Python, Visual Basic, Ruby, etc. The software project source codes are not compatible with later versions of Visual Studio. Repast Simphony is the latest version of Repast, which is combined with the powerful Eclipse integrated development environment, and also automated connectors to additional tools such as R, VisAD, MATLAB, Weka, and iReport (Getchell 2008).

2.15.11 *Agent Analyst Extension*

In this section, it is intended to present an overview of the Agent Analyst extension, which is an agent-based modelling and simulation extension for the ESRI-ArcGIS. Agent Analyst integrates the functionalities of the Repast simulation environment with the strengths and flexibilities of ArcObjects and ArcGIS in spatial analysis. Agent Analyst has the capability to integrate ABMS with GIS. GIS modellers are able to simulate environment behaviours and processes as change and movement over time by means of this extension (e.g. simulate land use and cover changes, predator–prey communications). This can help ABMS modellers to integrate detailed real-world biophysical data to execute complex spatial processes, as well as study how behaviour is constrained by space and geography. In addition, ABMS models can include update GIS data feeds for circumstances (e.g. fire-fighting, disaster management). This extension allows modellers to create, customise, and perform Repast models through the ArcGIS 9.2 geoprocessing framework, including access through the ArcToolbox, Model Builder, and ArcMap. Additionally, the Agent Analyst GUI allows users to create agents, schedule simulations, establish mappings by ArcGIS layers, and specify the behaviour and interactions of each agent (Bertelle et al. 2009).

2.16 Selection of ABM Implementation Toolkit

As it has been stated before, numerous toolkits have been developed in order to achieve ABM interests. Several software packages are also available for developing agent-based models that can facilitate the implementation process. For instance, simulation software often negates the necessity of developing agent-based models through low-level programming languages (e.g. Java, C++, Visual Basic, etc.). However, this brings some restrictions for developers to design their customised framework. As an example, some ABM software may support particular environments (e.g. either raster or vector), or agent transition rules might be limited in forms such as Von Neumann or Moore type. Furthermore, modellers will be constrained to the use of functions provided by the software, particularly while the toolkit has been written in its own programming engine (e.g. NetLogo) (Castle and Crooks 2006).

Each model has to be tested in their particular environments and, since the aim of this dissertation is to run a geosimulation prototype within ArcGIS software, it was decided to employ GIS functions for this simulation. In other words, it is intended to code an ABM environment within GIS software. Table 2.1 is a sample of open source simulation toolkits comparing the specified criteria and their advantages or disadvantages.

Agents can be designed and parameterised in a variety of ways, depending on the goals of the MAS (Valbuena et al. 2008; Robinson et al. 2007; Janssen and Ostrom 2006), for instance, agents can represent governmental regulation, land developers actions and residents behaviours at different environmental and financial circumstances (Valbuena et al. 2008; Ligtenberg et al. 2004; Monticino et al. 2007). As a matter of fact, the decision-making process has to be particularly parameterised by decision rules. These rules can be defined according to the expert knowledge or/and other researches' outcomes. Agent parameterisation with others' findings is the more frequent approach in MAS (Berger and Schreinemachers 2006; Valbuena et al. 2008). Alternatively, the parameterisation of agents with expert knowledge and empirical data facilitates understanding the real LUCC process. Most researches benefit from empirical data to specify and parameterise agents relying on huge data gathering (e.g. Valbuena et al. 2008; Jepsen et al. 2006; Castella et al. 2005; Huigen 2004; Bousquet et al. 2001).

2.17 Designing a Multi Agent System

In order to design a multi agent system, we will pursue the following stages, which are mentioned here:

Step 1: Collection and analysis of required input data for multiagents geosimulation,

Table 2.1 Comparison of open source simulation toolkits (Smith et al. 2007)

	Swarm	MASON	Repast
Developers	Santa Fe institute/Swarm development group, USA	Centre for social complexity, George Mason university, USA	University of Chicago, department of social science research computing, USA
Date of inception	1996	2003	2000
Implementation language	Objective-C/Java	Java	Java/Python/Microsoft .Net
Operating system	Windows, UNIX, Linux, Mac OSX	Windows, UNIX, Linux, Mac OSX	Windows, UNIX, Linux, Mac OSX
Required programming experience	Strong	Strong	Strong
Integrated GIS functionality	Yes	None	Yes (e.g. OpenMap, Java Topology Suite, and GeoTools)
Integrated charting/graphing/statistics	Yes (e.g. R and S-plus statistical packages)	None	Yes (e.g. Colt statistical package, and basic Repast functionality for simple network statistics)
Availability of demonstration models	Yes	Yes	Yes
Source code of demonstration models	Yes	Yes	Yes
Tutorials/how-to documentation	Yes	Yes	Yes

Step 2: Evaluation of existing geosimulation frameworks,
Step 3: Selection of an appropriate platform for geosimulation performance,
Step 4: Consideration of the land change drivers,
Step 5: Classification of agents,
Step 6: Specification of related factors and variables to each particular agent,
Step 7: Agents combination and reflection of agents' interactions,
Step 8: Creation and performance of multiagents geosimulation,
Step 9: Collection and analysis of the multiagents geosimulation results arising from the prototype,
Step 10: Verification and validation of multiagents geosimulation results,
Step 11: Report of the multiagents geosimulation' results,
Step 12: Visualisation of the multiagents geosimulation outcomes.

Figure 2.4 demonstrates a schematic workflow of geosimulation performance.

2.18 Fuzzy Decision Theory in Geographical Entities

Many phenomena exist with a degree of vagueness or uncertainty. Some terrestrial objects cannot be appropriately expressed with crisp sets. Kainz (2008) states that

> In human thinking and language we often use uncertain or vague concepts. Our thinking and language is not binary, i.e. black and white, zero or one, yes or no. In real life, we add much more variation to our judgments and classifications. These vague or uncertain concepts are said to be fuzzy. We encounter fuzziness almost everywhere in our everyday lives.

Fuzzy set theory was developed by Zadeh (1965) and extended by many authors, notably by Dubois and Prade (1979), to model uncertainties to allow for a more general theory of uncertainty than probability theory models. There is often confusion in the semantics of uncertainty pertaining to probability, interval, fuzzy, and possibility. GIS applications have seldom, if ever, used possibilistic geographical analysis. There are many reasons for this. Perhaps the most significant reason is that fuzzy set theory, as distinguished from possibility theory, is not always clear. Second, since geographical entities are often fuzzy (boundaries are gradual or transitional in nature between geographical entities) the use of possible entities is frequently omitted. Third, since Zadeh (1965) developed possibility theory via fuzzy set theory, most authors do not make a distinction and consider possibility distributions the same as fuzzy membership functions (Lodwick 2007).

Fuzzy set theory incorporates some concepts that can be used to overcome some of the problems described above, modelling some types of uncertainty associated to geographical information, as well as its heterogeneity. They enable the development of an alternative data model that integrates characteristics of the object and the field data models, where the geographical information is represented with fuzzy geographical entities, which are geographical entities (GE) represented

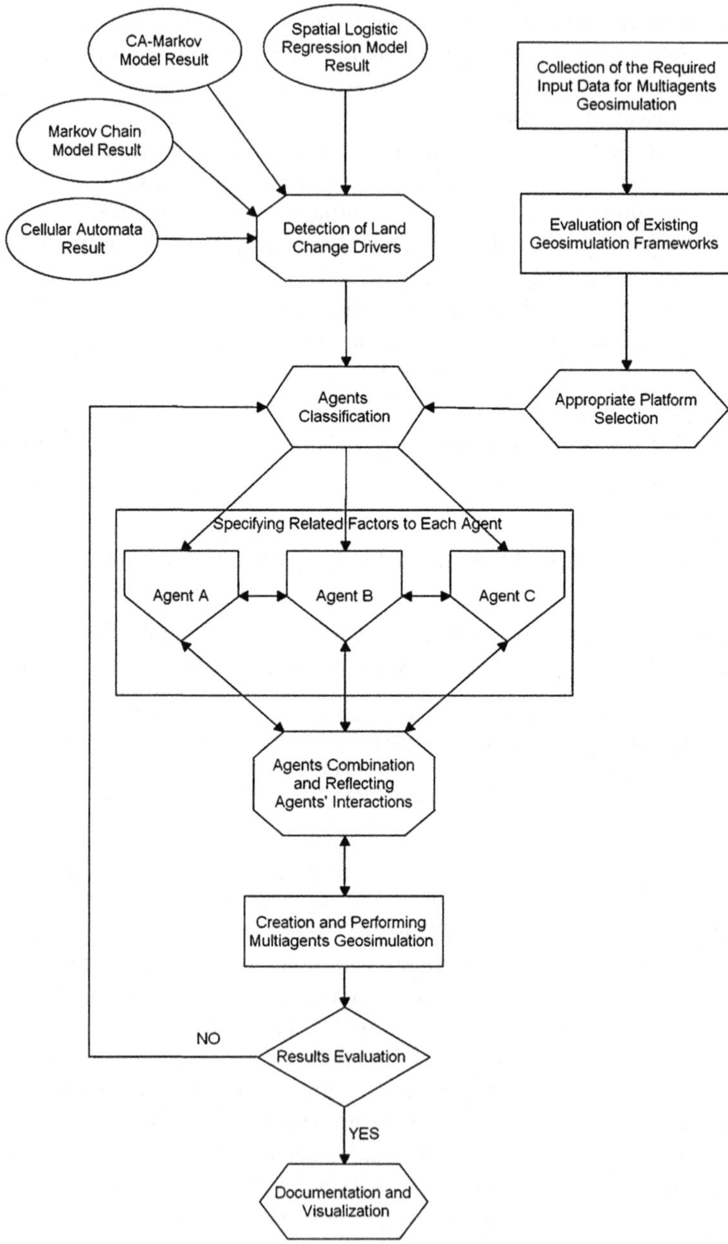

Fig. 2.4 Schematic view of multiagents geosimulation implementation

with surfaces. Using fuzzy geographical entities is an efficient way to represent the positional uncertainty of geographical entities, or a gradual variation between them, but their inclusion in geographical information systems requires not only their construction but also the development of operators capable of processing them (Lodwick 2007).

2.18.1 Fuzzy Geographical Entities

A geographic entity is characterised by an attribute and a geographical location. These two components of the geographic entities are intimately related, and therefore, the errors and uncertainty associated to each of them are also related. A fuzzy geographic entity (FGE), E_A, characterized by attribute "A", is a geographic entity whose position in the geographical space is defined by the fuzzy set:

$E_A = \{(x, y)$: (x, y) belongs to the GE characterised by attribute "A"$\}$

With a membership function, every location in the space of interest is defined between 0 and 1. The membership value 1 stands for full membership, and the membership value 0 represents no membership and values in between correspond to degree of membership to E_A, decreasing from 1 to 0. To model the uncertainty or errors regarding the positioning of an attribute, two cases may be considered. One is when the attribute is defined using only a concept such as buildings, forest, or rivers, and the other case is when the attribute definition is based on measurable quantities. In the first case, the difficulty to position the attributes on the ground depends on the details given on its definition and upon the heterogeneity of those attributes in the region under study. The details used in the attribute definition should be such that it is clear what the attribute represents, but note that if too many details are given, the identification of all those details in the ground may complicate the operator's work, since it may be difficult to identify, for example, in an aerial photograph, if a hut is made of wood or brick. So, the attribute definition should be adapted to the methods and sources of information available to identify the entities.

Whenever the operator has some difficulties in the classification, he may always assign a degree of uncertainty to the entity. For example, if there is some uncertainty whether a certain entity should be considered a building or not, a degree of uncertainty may be assigned to it. These degrees of uncertainty are subjective and only indicative in the sense that some difficulties in the classification were found. They are assigned to the entity as a whole, since it is an indivisible object. Note that, in this case, the grade of membership represents uncertainty in the attribute that should be assigned to the entire region. The outcome of this process is then a GE with a constant grade of membership to an attribute. These grades of membership translate degrees of membership to the attribute defining the GE and not uncertainty on the geographical space. They result from lack of data to assign the correct attribute or lack of attribute definition. In other situations, the uncertainty is not in the identification of the attribute corresponding to a certain GE, but in the

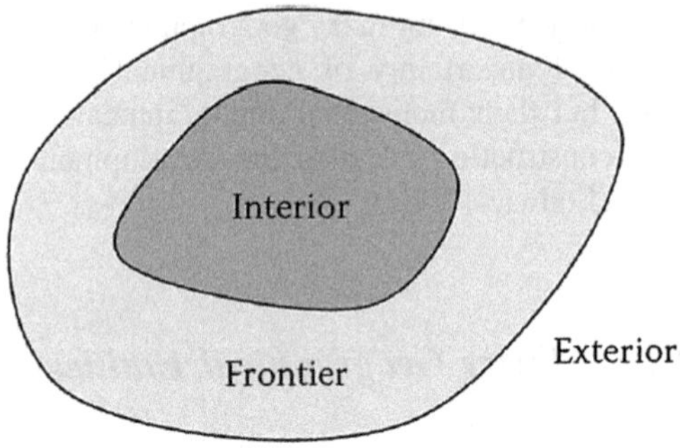

Fig. 2.5 Egg-Yolk approach to represent GEs (Cohn and Gotts 1996)

identification of its exact location. In this case, a fuzzy set may be used to express the entities' location in the geographical space.

This corresponds to the identification of the core of the fuzzy set and its support, and therefore all the region between the support and the core belongs to the uncertainty region. If it is possible to differentiate further inside the uncertainty region, then α-levels may be identified, if not, only an uncertainty region may be used.

This process generates fuzzy geographic entities, which correspond to the "egg-yolk" approach (Cohn and Gotts 1996). This approach considers that the geographic entities are formed by three regions: the interior, the frontier, and the exterior, where the frontier is represented not by lines, but by a region with any dimension or shape, and that may be considered homogeneous or heterogeneous (see Fig. 2.5). The "egg-yolk" representation is a simplified representation of fuzzy geographic entities and is convenient when the geographic entities are to be represented using the vector data structure and to establish neighbourhood relations between geographic entities with uncertain or fuzzy geographical position (Lodwick 2007).

2.18.2 Processing Fuzzy Geographical Entities

The use of fuzzy geographical entities in GIS environments requires operators capable of processing this type of entity. The instant approach to process FGEs is to convert them into crisp entities and use the usual operators to perform the necessary operations. Since the α-levels of fuzzy sets are crisp sets, the easiest way to convert fuzzy GEs to crisp GEs is to substitute the entity by one of its α-levels. By means of this approach, a variety of versions of the FGE may be achieved according to the needs of each application, selecting different α-levels to represent it. In this research, we will use fuzzy sets to represent prediction process of land use classes over time (Lodwick 2007).

2.19 The Analytical Hierarchy Process Weighting

The analytic hierarchy process (AHP) is a very pervasive and commonly used application for decision-making matters (De Feo and De Gisi 2010). Indeed, the AHP was developed by Thomas L. Saaty in the 1970s. The AHP affords a comprehensive rational structure in order to solve the decision-making problems, as well as characterising and quantifying its elements and conduction of the related elements towards overall goals, plus evaluating alternative solutions (De Feo and De Gisi 2010; Forman and Selly 2001; Saaty 1977).

In fact, the AHP has special advantages in multi-variable evaluation. It has been utilised in various research fields, such as natural science, economy and society (Ramanathan and Ganesh 1995). AHP is also becoming a common tool of eco-environment quality evaluation, for ecological environment is a large and multi-layered system (Hill et al. 2005; Klungboonkrong and Taylor 1998; Li et al. 2007; Yedla and Shrestha 2003). GIS-based AHP is popular because of its strong capacity to integrate various types of heterogeneous variables and its simplicity to obtain the weights of appropriate variables. Therefore, this reasoning promotes its strength in criteria weighting (Hossain and Das 2010; Tiwari et al. 1999).

The AHP method splits a complex multi-criteria decision matter into a hierarchy and performs on the basis of a pair-wise comparison of the importance of different criteria and sub-criteria (De Feo and De Gisi 2010; Forman and Selly 2001; Saaty 1977). According to Saaty (1977) the AHP process is based on three main steps. The first step is to establish a hierarchical structure, where the first hierarchy of a structure is the goal. The final hierarchy deals with identifying alternatives, while the middle hierarchy levels appraise certain factors or conditions.

The second step computes the element weights of various hierarchies by means of three sub-steps. The first sub-step creates the pair-wise comparison matrix, then a pair-wise comparison is conducted for each element based on an element of the upper hierarchy that is an evaluation standard. The second sub-step computes the eigenvalue and eigenvector of the pair-wise comparison matrix. The third sub-steps execute the consistency test. The difference between the dominant eigenvalue of the pair-wise comparison matrix ($\lambda_{\max}$) and the matrix dimension (k) is used in defining the inconsistency index, II (Hsu et al. 2008; Karamouz et al. 2007; Saaty 1999):

$$II = [(\lambda_{\max} - k)/(k - 1)] \tag{2.4}$$

Moreover, the inconsistency ratio (IR) is defined as (Hsu et al. 2008; Karamouz et al. 2007; Saaty 1999):

$$IR = II/CRI \tag{2.5}$$

The CRI presents the inconsistency index of the random matrix retrieved by calculating II for a randomly filled matrix. If IR <10%, then the consistency criterion is acceptable. Otherwise, the decision-maker has to refine the pair-wise comparisons. This procedure goes on until all the pair-wise comparisons satisfy the consistency criterion. The eigenvector of the pair-wise comparison matrix is used to estimate the relative weight of different choices. Finally, the third step of the AHP technique calculates the entire hierarchical weight. In reality, AHP generates an overall ranking of the solutions using the comparison matrix among the alternatives and information on the ranking of the criteria. Thus, the option with the highest eigenvector value is approved to be the first choice (Hsu et al. 2008; Karamouz et al. 2007; Saaty 1999).

In fact, the strength of AHP is that it allows for the creation of inconsistent relationships, besides affording the CR index as an indicator of the degree of consistency or inconsistency (Forman and Selly 2001). Thus, the AHP execution in this thesis will incorporate an opportunity to let the user define a satisfactory CR threshold value. In this thesis, it is intended to improve the performance of agent-based modelling by means of a combination of GIS, AHP and ABM; therefore, all socio-economic and environmental variables are combined according to their weights.

2.20 Moran's Autocorrelation Coefficient Analysis

Moran's autocorrelation coefficient denoted as Moran's-I is an extension of Pearson product-moment correlation coefficient to a univariate series. Recall that Pearson's correlation symbolised as ρ between two variables x and y both of extent n is:

$$\rho = \frac{\sum_{i=1}^{n}(x_i - \bar{x})}{\left[\sum_{i=1}^{n}(x_i - \bar{x})^2 \sum_{i=1}^{n}(y_i - \bar{y})^2\right]^{1/2}} \tag{2.6}$$

Where $\bar{x}$ and $\bar{y}$ are the sample means of both variables. ρ measures whether, on average, x_i and y_i are associated. For a single variable, for instance x, I will measure whether x_i and x_j with $i \neq j$, are associated. Note that with ρ x_i and x_j are not associated since the pairs (x_i, y_i) are assumed to be independent of each other.

Moran's I formula is as following:

$$I = \frac{n}{s_0} \frac{\sum_{i=1}^{n}\sum_{j=1}^{n}\omega_{ij}(x_i - \bar{x})(x_j - \bar{x})}{\sum_{i=1}^{n}(xi - \bar{x})^2} \tag{2.7}$$

Where ω_{ij} is the weight between observation i and j, and s_0 is the sum of all ω_{ij}:

$$S_0 = \sum_{i=1}^{n}\sum_{j=1}^{n}\omega_{ij} \tag{2.8}$$

The expected value of I under the null hypothesis of no autocorrelation is not equivalent to zero but given by $I_0 = -1/(n-1)$. The expected variance of I_0 is also known, and so we can make a test of the null hypothesis. If the observed value of I denoted $\hat{I}$ is significantly greater than I_0, then values of x are positively auto correlated, while if $\hat{I} < I_0$, this points out negative autocorrelation. This allows designing one or two-tailed tests in the standard way (Ellingson and Andersen 2002).

2.21 Accuracy Assessment and Uncertainty in Maps Comparison

Land use/cover change simulation models basically examine land change maps at two separated periods (t_0 and t_1) and then, by evaluation of the occurred changes within these two periods and change factors, an appropriate simulation model is performed. This performance will predict land change maps for future periods (i.e. t_2). This predicted map at point t_2 can be typically compared to a reference map (i.e. the map of reality) to estimate the model performance. Therefore, in case the result of this comparison demonstrates a high degree of similarity, then it can be proved that the model was successful to simulate the changes. Even though, this result cannot necessarily indicate that the model provided supplementary findings beyond what the scientist would have predicted without the model. As we examined in this research, and as some scientists believe for most of the LUCC models, the agreement between the reference map of t_1 and t_2 always appears to be better than the agreement between the predicted map of t_2 and its reference map, at the resolution at which the model was run. Hence, this causes alarm in the LUCC modelling amongst scientists.

2.21.1 Calibration and Validation

In this section, it is intended to clarify the distinction between two terms of 'Calibration' and 'Validation'. It is better to distinguish between these two terms, whereas, they will be used after the model execution. Nevertheless, in most of the read papers, it was complicated to note their distinctions.

Rykiel (1996) noted that

> Calibration is the estimation and adjustment of the model parameters and constraints to improve the agreement between model output and a data set,

whereas validation is

> a demonstration that a model within its domain of applicability possesses a satisfactory range of accuracy consistent with the intended application of the model.

Besides, in some cases, the terms of "modelled" and "simulated" were used in error. For instance, the word "modelled" means that the model was fitted and the term "simulated" indicated that something was predicted. Furthermore, in some other cases, there is a lack of methodology reflection (Wu and Webster 1998).

These issues lead scientific communities to a misunderstanding of the model's certainty. Two separated types of data should be utilised for the calibration and evaluation processes. There are some ways to break up the calibration data from the validation data. Separation through time is one of the usual ways. Hence, if the model aims to predict the change on the landscape after time t_1, then any information at t_1 or before t_1 is justifiable to use for calibration. For instance, a usual calibration method is to carry out statistical regression on the change quantity between point t_0 and t_1. The results of the regression are fitted estimates. The fitted parameters and the regression relationship might be used to extrapolate the change between point t_1 and t_2; thus, any information subsequent to time t_1 cannot be used in the calibration process. The validation process compares the predicted map of time t_2 to the reference map of time t_2. Separation through space is another general method to separate calibration information from validation information. In this scheme, the model uses data from one study site to fit the parameters, and then the fitted model is applied to a different site to predict change. Distinction between the calibration process and the validation process can help to guarantee that the model is not over-fitted (Pontius et al. 2004).

Before implementing the simulation process, it has to compare the resultant maps arising from modelling execution to ensure the validity of the model. In fact, there is no universal concord to evaluate the goodness-of-fit of validation (Rykiel 1996). Each particular model comprises a specific purpose, and hence, the criterion should be related to the purpose. Besides, scale is essential to consider throughout any comparison of maps, since results might be sensitive to scale and, also, certain patterns may be evident at only certain scales (Quattrochi and Goodchild 1997; Kok et al. 2001; Pontius et al. 2004).

2.21.2 Techniques of Validation for Land Change Models

According to the definition of the United States Geological Survey (USGS), regarding the accuracy of geospatial data;

> The closeness of results of observations, computations, or estimates to the true values or the values accepted as being true.

Nonetheless, the term "truth" has a certain definition. Accuracy assessment is one of the most imperative and significant steps in remote sensing and map analysis. GIS and remote sensing outcomes are being used as basic inputs for local, national, and global decisions; therefore, precise and accurate outputs lead researches towards correct routes. New users have to be taught about the reliability of the maps which are produced from GIS and remote sensing tasks (Banko 1998).

2.21.2.1 Visual Interpretation

A visual interpretation can give the scientists a better general assessment of model performance. Pure visual comparison is vulnerable to the personal opinion of the user; therefore, one scientist might believe the results perfect, and another might identify them poor.

2.21.2.2 Kappa Coefficient

Statistical techniques of results comparison are helpful to discover patterns that the individual mind ignores and also to facilitate communication between scientists. The Kappa coefficient was set up to the remote sensing societies in the early 1980s by Congalton et al. (1983), Congalton and Mead (1983), Cohen (1960) and became a pervasive measurement index for classification accuracy (Huang et al. 2002). It was recommended as a standard by Rosenfield and Fitzpatrick-Lins (1986). The Kappa coefficient measures the overall agreement of a matrix. The ratio of the summation of diagonal values to the total number of pixel counts in the matrix, the Kappa coefficient considers non-diagonal elements as well (Rosenfield and Fitzpatrick-Lins 1986). In fact, the Kappa coefficient computes the fraction of agreement after elimination of the chance agreement from considerations. A Kappa of 0 arises while the agreement between actual and reference maps equals chance agreement, and Kappa increases up to 1(Banko 1998; Lakide 2009).

2.21.2.3 Relative Operating Characteristic

Pontius et al. (2004) have suggested a comparison technique that considers the agreement between two pairs of maps. The first comparison performs between the reference map of point t_1 and the reference map of time t_2. The second comparison performs between the predicted map of point t_2 and the reference map of point t_2. Eventually, the procedure evaluates the first comparison in comparison with the second comparison.

Swets (1986, 1988) depicts the logic of the ROC, although other researchers in this field explain how to compute the ROC in the digital maps comparison (Pontius et al. 2004; Pontius and Batchu 2003). The relative operating characteristic is a statistical measurement to compare a Boolean variable versus a categorical variable. The ROC is able to compute the accuracy of the prediction at several diverse threshold levels. For each threshold domain, each cell of probability surface map is reclassified as either over or under the threshold (Pontius and Batchu 2003).

Pontius and Batchu (2003) believe that the ROC is an outstanding method for analysing propensity surface values. Moreover, the ROC compares two maps specification in two separate ways: in terms of location and also quantity. ROC achieves this by the goodness-of-fit calculation of the validation at various thresholds, thereafter, aggregating the information at all thresholds into one measure of agreement. Accordingly, this method is purported to have distinct values concerning the goodness-of-fit of location versus the goodness-of-fit of quantity; therefore, modellers will be able to improve the predictive model capability.

2.22 Summary

This chapter started with a brief introduction about the chapter and then introduced land use and land cover terms (Sect. 2.2) before making the distinction between these two terms, thereby highlighting the LUCC causes and consequences (Sect. 2.3). LUCC driving forces (Sect. 2.4) and LUCC simulation (Sect. 2.5) were introduced thereafter. The typical methodologies for trend evaluation of land use change was also discussed (Sect. 2.6). Section 2.7 dealt with how to predict the upcoming land use patterns and Sect. 2.8 with how to simulate urban sprawl. A comprehensive discussion over the popular and existing approaches in LUCC studies was presented in Sect. 2.8. A comprehensive explanation of geosimulation methodology, its terminology and characteristics was also depicted (Sects. 2.10–2.12). Later, Sects. 2.13, 2.14 argued about the differentiation between cellular automata and the GAS model. The existing simulation environments were taken into consideration to pick the optimum system for this research (Sect. 2.15). The fundamentals of a fuzzy decision system, the AHP weighting system, and the Moran autocorrelation coefficient were explained (Sect. 2.16–2.20). The usual map comparison methods were explained in order to evaluate the certainty of output maps.

References

Anylogic (2006) Anylogic. Available at: http://www.xjtek.com/

Bakker MM, van Doorn AM (2009) Farmer-specific relationships between land use change and landscape factors: introducing agents in empirical land use modelling. Land Use Policy 26(3):809–817

Bandini S, Manzoni S, Vizzari G (2009) Agent based modeling and simulation: an informatics perspective. J Artif Soc Social Simul 12:4

Banko G (1998) A review of assessing the accuracy of classifications of remotely sensed data and of methods including remote sensing data in forest inventory, Working Papers ir98081, International Institute for Applied Systems Analysis: Austria

Batty M (2005) Cities and complexity: understanding cities with cellular automata, agent-based models, and fractals. The MIT Press, Cambridge

Batty M, Jiang B (1999) Multi-agent simulation new: approaches to exploring space-time dynamics in GIS. Centre for Advanced Spatial Analysis (UCL), London, UK

Bazghandi A, Pouyan A (2008) Considering geographic information systems in buyer/seller agents simulation. In: information and communication technologies: from theory to applications, 2008. ICTTA 2008. 3rd international conference on, pp 1–5

Benenson I, Torrens PM (2003) Geographic automata systems: a new paradigm for integrating GIS and geographic simulation. In: Martin D (ed) Proceedings of the 7th international conference on geocomputation, Southampton, GeoComputation 2003 CD-ROM

Benenson I, Torrens PM (2004) Geosimulation: automata-based modeling of urban phenomena. Wiley, New York

Benenson I, Aronovich S, Noam S (2005) Let's talk objects: generic methodology for urban high-resolution simulation. Comput Environ Urban Syst 29(4):425–453

Berger T (2001) Agent-based spatial models applied to agriculture: a simulation tool for technology diffusion, resource use changes and policy analysis. Agric Econ 25(2–3):245–260

Berger T, Schreinemachers P (2006) Creating agents and landscapes for multiagent systems from random samples. Ecol Soc 11(2):19

Bertelle C, Duchamp GHE, Kadri-Dahmani H (2009) Complex systems and self-organization modelling. Springer, New York

Bonabeau E (2002) Agent-based modeling: methods and techniques for simulating human systems. Proc Natl Acad Sci U S A 99(90003):7280–7287

Bousquet F, Le Page C (2004) Multi-agent simulations and ecosystem management: a review. Ecol Modell 176(3–4):313–332

Bousquet F, Le Page C, Bakam I, Takforyan A (2001) Multiagent simulations of hunting wild meat in a village in eastern Cameroon. Ecol Modell 138(1–3):331–346

Brail RK, Klosterman RE (2001) Planning support systems: integrating geographic information systems, models, and visualization tools. ESRI Inc., New York

Camazine S, Deneubourg J, Franks N, Sneyd J, Theraulaz G, Bonabeau E (2003) Self-organization in biological systems. Princeton University Press, Princeton

Castella J, Boissau S, Trung T, Quang D (2005) Agrarian transition and lowland-upland interactions in mountain areas in northern Vietnam: application of a multi-agent simulation model. Agric Syst 86(3):312–332

Castle CJ, Crooks AT (2006) Principles and concepts of agent-based modelling for developing geospatial simulations, centre for advanced spatial analysis (UCL). UCL (University College London), London

Cohen J (1960) A coefficient of agreement for nominal scales. Educ Psychol Measur 20(1):37–46

Cohn AG, Gotts NM (1996) The 'Egg-Yolk' representation of regions with indeterminate boundaries. In: Burrough PA, Frank AU (eds) Geographic objects with indeterminate boundaries. Taylor and Francis, London, pp 171–187

Congalton R, Mead R (1983) A quantitative method to test for consistency and correctness in photointerpretation. Photogramm Eng Remote Sens 49(1):69–74

Congalton RG, Oderwald RG, Mead RA (1983) Assessing landsat classification accuracy using discrete multivariate analysis statistical techniques. Photogramm Eng Remote Sens 49:1671–1678

Crawford TW, Messina JP, Manson SM, O'Sullivan D (2005) Complexity science, complex systems, and land-use research. Env Planning B 32:792–798

Crooks AT (2006) Exploring cities using agent-based models and GIS. Proceedings of the agent 2006 conference on social agents: results and prospects, university of Chicago and Argonne national laboratory, Chicago, IL, USA

Crooks AT (2007a) The repast simulation/modelling system for geospatial simulation, centre for advanced spatial analysis (University College London): Working Paper 123, London, UK

Crooks AT (2007b) Experimenting with cities: utilizing agent-based models and GIS to explore urban dynamics. University College London, London

Crooks AT, Castle C, Batty M (2008) Key challenges in agent-based modelling for geo-spatial simulation. Comput Env Urban Syst 32(6):417–430

De Feo G, De Gisi S (2010) Using an innovative criteria weighting tool for stakeholders involvement to rank MSW facility sites with the AHP, waste management, vol 30, issue 11. Special thematic section: sanitary land filling, pp 2370–2382

Dubois D, Prade H (1979) Fuzzy real algebra: some results. Fuzzy Sets Syst 2(4):327–348

Ducheyne E (2003) Multiple objective forest management using GIS and genetic optimisation techniques, PhD thesis, faculty of agricultural and applied biological sciences. University of Ghent, Belgium

Ellingson AR, Andersen DC (2002) Spatial correlations of diceroprocta apache and its host plants: evidence for a negative impact from tamarix invasion. Ecol Entomol 27(1):16–24

Ellis E, Pontius Jr RG (2006) land-use and land-cover change—encyclopedia of earth. Available at: http://www.eoearth.org/article/land-use_and_land-cover_change

Epstein JM (2007) Agent-based computational models and generative social science, in generative social science studies in agent-based computational modeling. Princeton University Press, Princeton, pp 41–60

Epstein JM, Axtell RL (1996) Growing artificial societies: social science from the bottom up, 1st edn. The MIT Press, Cambridge

Ettema D, De Jong K, Timmermans H, Bakema A (2007) PUMA: multi-agent modelling of urban systems. In modelling land-use change. The geojournal library. Springer, Netherlands, pp 237–258

Forman EH, Selly MA (2001) Decision by objectives: how to convince others that you are right. World Scientific, Singapore

Franklin S, Graesser A (1996) Is it an agent, or just a program?: a taxonomy for autonomous agents. In: Müller JP, Wooldridge MJ, Jennings NR (eds) Proceedings of the third international workshop on agent theories, architectures, and languages, Springer, pp 21–35

Getchell A (2008) Agent-based modeling, university of California, Davis. Available at: http://www2.econ.iastate.edu/tesfatsi/AgentBasedModeling.AdamGetchell.phy250.Report.pdf

Hill MJ, Braaten R (2005) Multi-criteria decision analysis in spatial decision support: the ASSESS analytic hierarchy process and the role of quantitative methods and spatially explicit analysis. Environ Model Softw 20(7):955–976

Holland J (1996) Hidden order: how adaptation builds complexity, 1st edn. Addison Wesley Longman, Redwood City

Hossain MS, Das NG (2010) GIS-based multi-criteria evaluation to land suitability modelling for giant prawn macrobrachium rosenbergii farming in companigonj upazila of noakhali Bangladesh. Comput Electron Agric 70(1):172–186

Hsu P, Wu C, Li Y (2008) Selection of infectious medical waste disposal firms by using the analytic hierarchy process and sensitivity analysis. Waste Manage 28(8):1386–1394

Huang C, Davis LS, Townshend JRG (2002) An assessment of support vector machines for land cover classification. Int J Remote Sens 23(4):725–749

Huigen MGA (2004) First principles of the MameLuke multi-actor modelling framework for land use change, illustrated with a Philippine case study. J Environ Manage 72(1–2):5–21

Irwin EG, Bockstael NE (2002) Interacting agents, spatial externalities and the evolution of residential land use patterns. J Econ Geogr 2(1):31–54

Irwin EG, Geoghegan J (2001) Theory, data, methods: developing spatially explicit economic models of land use change. Agric Ecosyst Env 85(1–3):7–24

Janssen MA, Ostrom E (2006) Empirically based, agent-based models. Ecol Soc 11(2):37

Jepsen MR, Leisz S, Rasmussen K, Jakobsen J, Müller-Jensen L, Christiansen L (2006) Agent-based modelling of shifting cultivation field patterns, Vietnam. Int J Geog Inf Sci 20(9):1067–1085

Kainz W (2008) Fuzzy logic and GIS. University of Vienna, Available at: http://homepage.univie.ac.at/wolfgang.kainz/Lehrveranstaltungen/ESRI_Fuzzy_Logic/File_2_Kainz_Text.pdf

Karamous M, Zahraie B, Kerachian R, Jaafarzadeh N, Mahjouri N (2007) Developing a master plan for hospital solid waste management: a case study. Waste Manage 27(5):626–638

Klungboonkrong P, Taylor MAP (1998) A microcomputer-based- system for multicriteria environmental impacts evaluation of urban road networks. Comput Env Urban Syst 22(5):425–446

Kok K, Veldkamp A (2001) Evaluating impact of spatial scales on land use pattern analysis in central America. Agric Ecosyst Env 85(1–3):205–221

Kok K, Farrow A, Veldkamp A, Verburg PH (2001) A method and application of multi-scale validation in spatial land use models. Agric Ecosyst Env 85(1–3):223–238

Koomen E, Stillwell J, Bakema A, Scholten HJ (2007) Modelling land-use change: progress and applications. Springer, New York

Lakide V (2009) Classification of synthetic aperture radar images using particle swarm optimization technique. MSc. thesis, National Institute of Technology Rourkela. Available at: http://ethesis.nitrkl.ac.in/1438/

Lambin EF, Geist HJ, Ellis E (2007) Causes of land-use and land-cover change. In encyclopedia of earth

Levy S (1992) Artificial life. The quest for a new creation. Penguin

Li X, Liu X (2007) Defining agents' behaviors to simulate complex residential development using multicriteria evaluation. J Environ Manage 85(4):1063–1075

Ligtenberg A, Wachowicz M, Bregt AK, Beulens A, Kettenis D (2004) A design and application of a multi-agent system for simulation of multi-actor spatial planning. J Environ Manage 72(1–2):43–55

Lodwick W (2007) Fuzzy surfaces in GIS and geographical analysis. CRC Press, Boca Raton

Longley P, Batty M (2003) Advanced spatial analysis: the CASA book of GIS. ESRI Inc, California

Macal CM, North MJ (2005) Tutorial on agent-based modeling and simulation. In: Kuhl ME, Steiger NM, Armstrong FB, Joines JA (eds) Proceedings of the 37th conference on winter simulation, Winter Simulation Conference, Orlando, Florida, pp 2–15

Matthews R (2006) The people and landscape model (PALM): towards full integration of human decision-making and biophysical simulation models. Ecol Modell 194(4):329–343

Meyer WB, Turner BL (1994) Changes in land use and land cover: a global perspective. Cambridge University Press, Cambridge

Monticino M, Acevedo M, Callicott B, Cogdill T, Ji M, Lindquist C (2007) Coupled human and natural systems: a multi-agent-based approach. Environ Model Softw 22(5):656–663

Macal CM, North MJ (2006) Tutorial on agent-based modeling and simulation part 2: how to model with agents. In: Perrone LF, Wieland FP, Liu J, Lawson BG, Nicol DM, Fujimoto RM (eds) Proceedings of the 38th conference on winter simulation, Winter Simulation Conference, Monterey, California, pp 73–83

Openshaw S (1983) The modifiable areal unit problem concepts and techniques in modern geography, 28th edn. Geo Books, Norwich

O'Sullivan D, Macgill JR, Yu C (2003) Agent-based residential segregation: a hierarchically structured spatial model. Proceedings of agent 2003 conference on challenges in social simulation, The University of Chicago, Chicago, pp 493–507

Parker DC, Manson SM, Janssen MA, Hoffmann MJ, Deadman P (2003) Multi-agent systems for the simulation of land-use and land-cover change: a review. Ann Assoc Am Geogr 93:314–337

Pontius RG Jr, Batchu K (2003) Using the relative operating characteristic to quantify certainty in prediction of location of land cover change in India. Trans GIS 7(4):467–484

Pontius RG Jr, Chen H (2006) GEOMOD modeling, idrisi andes help contents. Clark University, Massachusetts

Pontius RG Jr, Schneider LC (2001) Land-cover change model validation by an ROC method for the ipswich watershed, Massachusetts, USA. Agric Ecosyst Environ 85(1–3):239–248

Pontius RG Jr, Cornell JD, Hall CAS (2001) Modeling the spatial pattern of land-use change with GEOMOD2: application and validation for costa rica. Agric Ecosyst Env 85(1–3):191–203

Pontius RG Jr, Huffaker D, Denman K (2004) Useful techniques of validation for spatially explicit land-change models. Ecol Modell 179(4):445–461

Quattrochi DA, Goodchild MF (1997) Scale in remote sensing and GIS, 1st edn. CRC Press, Boca Raton

Ramanathan R, Ganesh LS (1995) Energy resource allocation incorporating qualitative and quantitative criteria: an integrated model using goal programming and AHP. Socio-Econ Planning Sci 29(3):197–218

Repenning A, Ioannidou A, Zola J (2000) Agentsheets: end-user programmable simulations. J Artif Soc Social Simul 3:3

Rindfuss RR, Walsh SJ, TurnerII BL, Fox J, Mishra V (2004) Developing a science of land change: challenges and methodological issues. Proc Natl Acad Sci U S A 101(39): 13976–13981

Robinson DT, Brown DG, Parker DC, Schreinemachers P, Janssen MA, Huigen M, Wittmer H, Grotts N, Promburom P, Irwin E, Berger T, Gatzweiler F, Barnaud C (2007) Comparison of

empirical methods for building agent-based models in land use science. J Land Use Sci 2(1):31–55
Rosenfield GH, Fitzpatrick-Lins K (1986) A coefficient of agreement as a measure of thematic classification accuracy. Photogramm Eng Remote Sens 52(2):223–227
Russell S, Norvig P (2009) Artificial intelligence: a modern approach, 3rd edn. Prentice Hall, Englewood Cliffs
Rykiel EJ (1996) Testing ecological models: the meaning of validation. Ecol Modell 90(3):229–244
Saaty TL (1977) A scaling method for priorities in hierarchical structures. J Math Psychol 15(3):234–281
Saaty TL (1999) Decision making for leaders: the analytic hierarchy process for decisions in a complex world, New edition 2001, 3rd edn. RWS Publications, Pittsburgh
Sawyer RK (2003) Artificial societies: multiagent systems and the micro-macro link in sociological theory. Sociol Methods Res 31:325–363
Showalter P, Lu Y (2009) Geospatial techniques in Urban hazard and disaster analysis. Springer, The Netherlands
Smith MJD, Goodchild MF, Longley PA (2007) Geospatial analysis: a comprehensive guide to principles, techniques and software tools, 2nd edn. Troubador Publishing Ltd, Kibworth
Swets JA (1986) Indices of discrimination or diagnostic accuracy: their ROCs and implied models. Psychol Bull 99(1):100–117
Swets JA (1988) Measuring the accuracy of diagnostic systems. Science (New York) 240(4857):1285–1293
Timmermans H (2003) The saga of integrated land use-transport modeling: how many more dreams before we wake up. In Keynote paper auf der 10th international conference on travel behavior research, Luzern
Tisue S, Wilensky U (2004) NetLogo: a simple environment for modelling complexity. International conference on complex systems (ICCS 2004), Boston, MA, pp 16–21
Tiwari DN, Loof R, Paudyal GN (1999) Environmental-economic decision-making in lowland irrigated agriculture using multi-criteria analysis techniques. Agric Syst 60(2):99–112
Tobias R, Hofmann C (2004) Evaluation of free Java-libraries for social-scientific agent based simulation. J Artif Soc Social Simul 7:1
Torrens P (2006a) Simulating sprawl. Ann Assoc Am Geogr 96(2):248–275
Torrens PM (2006b) Geosimulation and its application to Urban growth modeling. Springer, London, pp 119–134
TurnerII BL, Skole D, Sanderson S, Fischer G, Fresco L, Leemans R (1995) Land-use and land-cover change science/research plan, IGBP report no. 35, HDP report no. 7, Stockholm and Geneva
Valbuena D, Verburg PH, Bregt AK (2008) A method to define a typology for agent-based analysis in regional land-use research. Agric Ecosyst Environ 128(1–2):27–36
Veldkamp A, Lambin EF (2001) Predicting land-use change. Agric Ecosyst Env 85(1–3):1–6
Verburg PH, Schot P, Dijst M, Veldkamp A (2004) Land use change modelling: current practice and research priorities. GeoJournal 61(4):309–324
Wooldridge MJ, Jennings NR (1995) Intelligent agents: theory and practice. Knowl Eng Rev 10(2):115–152
Wu F, Webster CJ (1998) Simulation of land development through the integration of cellular automata and multicriteria evaluation. Env Plan B 25(1):103–126
Yang Q, Li X, Shi X (2008) Cellular automata for simulating land use changes based on support vector machines. Comput Geosci 34(6):592–602
Yedla S, Shrestha R (2003) Multi-criteria approach for the selection of alternative options for environmentally sustainable transport system in Delhi. Transp Res Part A 37:717–729
Zadeh L (1965) Fuzzy sets. Inf Control 8(3):338–353

Chapter 3
Study Area Description

3.1 Introduction

In this chapter, the aim is to give a brief explanation about the case study which has been selected to accomplish this research. It has been assumed that Tehran, the capital of Iran, is an ideal case study area, because it has been seriously afflicted with urban sprawl. This problem is so great as to be devouring the surrounding farming lands as well as open lands and, consequently, turning them into urban, built-up areas. Moreover, this research will integrate affordable spatial explicit data and non-spatial information in the field of socioeconomic, environmental and other affiliated variables.

3.2 Case Study Description

It was considered vital to choose an appropriate area in order to implement an ABM and geosimulation to obtain enhanced and significant outcomes. Hence, it was reasoned that the Tehran metropolitan area can serve as a match case for this research, because of the severity and extent of urban expansion in the metropolis.

Tehran, as the capital of Iran, is one of the few capitals in the world which is not located beside a river or even close to the sea. Mountains predominate and surround the city from the north and east (Fardi 2010). The selected area for this research includes Tehran city as well as its suburbs; in other words, approximately a Landsat image coverage of 164/35 row-path lattice. This area covers approximately 188,000 hectares and is shown in Fig. 3.1.

Within 200 years, decreasing mortality rates and an influx of migrants have altered Tehran from a 7.5 km^2 city of 14,500 residents into a mega-city of almost ten million people, expanding to an area of 620 km^2. It has subsumed flat and open lands to the east and west and 70 villages on its adjacent mountain slopes. Today's

J. Jokar Arsanjani, *Dynamic Land-Use/Cover Change Simulation: Geosimulation and Multi Agent-Based Modelling*, Springer Theses,
DOI: 10.1007/978-3-642-23705-8_3,

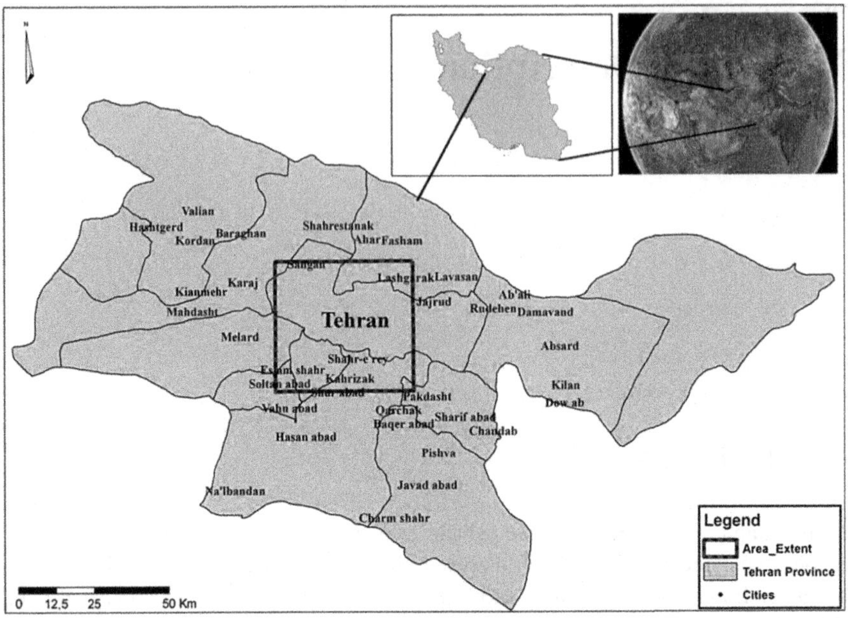

Fig. 3.1 Geographical extent of the study area, Tehran metropolitan area

Tehran varies 800 m in elevation from south to north, and the city's topographical variation is paralleled by the marked differences in class and the lifestyle of its inhabitants (Shahshahani 2003).

Tehran province comprises one of 31 provinces in Iran. It covers an extent of 18,910 km^2 and is located to the north of the central plateau of country. Tehran province borders Mazandaran province in the north, Qom province in the south, Semnan province in the east and Karaj province in the west. The metropolis of Tehran is the capital of the province as well as of Iran. As of June 2005, this province includes 13 townships, 43 municipalities, and 1,358 villages. Tehran metropolis, with an area of around 780 km^2, is the most heavily populated and biggest city in Iran. The population of its municipality has grown from 0.7 million in 1941 to approximately 7,230,000 in 2005. When including the surrounding areas and the commuting workforce, the metropolitan area now exceeds 12 million inhabitants. The population growth rate of Tehran was shown to be 0.6% between 2001 and 2005 (Demography Information 2006).

The metropolitan area of Tehran is surrounded on the northern and eastern sides by the Alborz Mountains, one of the highest mountain ranges in Iran with peaks above 5,670 m. Average elevation of the city is around 1,300 m above sea level, which increases towards the north. The city (the municipality of Tehran) covers an area of approximately 22 km north–south and around 35 km east–west—embedded in a 60 by 60 km domain. The city area is divided into 22 municipality districts (Iran Chamber Society 2001) which is demonstrated in Fig. 3.2.

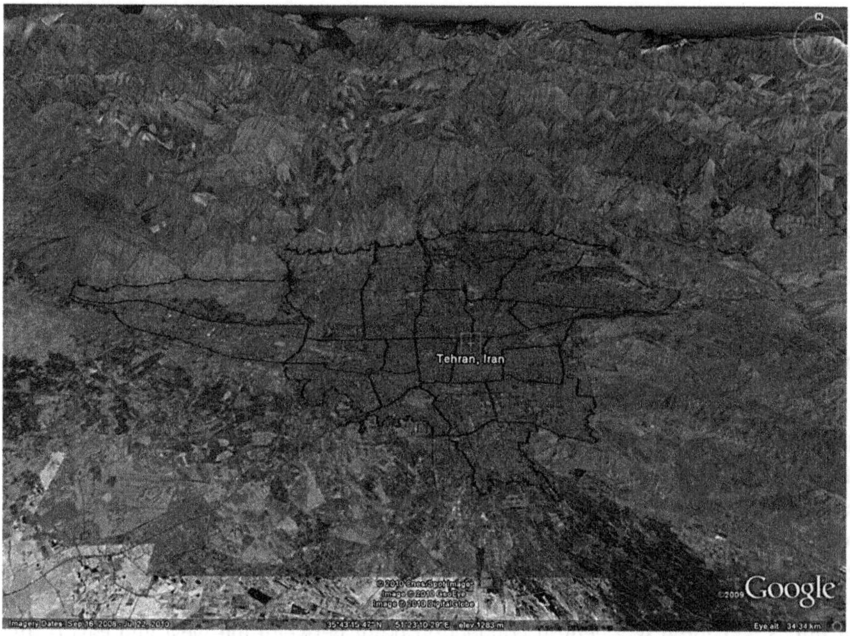

Fig. 3.2 Arial view of Tehran and its current districts

3.3 Geography

According to the Iranian statistics center of 2006 (the latest recorded demographic data), the province of Tehran consists of 13 counties and has more than 12.4 million inhabitants. Tehran is the most densely populated state of Iran. Approximately 86.5% of inhabitants live in urban areas and the remainder reside in the rural areas of the province (Demography Information 2006). The highest peak of the province is the Damavand mountain, with an elevation of more than 5,670 m above sea level, with the Varamin plains being the lowest point of the province, at around 790 m above sea level. The province includes more than 330 km^2 of forests, and over 12,800 km^2 of pasture land. The major rivers of this province are the Karaj River and the Jajrud River (Iran Chamber Society 2001). Tehran is surrounded by various mountains such as the Alborz mountains in the north; Savad-Kooh and Firooz-Kooh are situated in the northeast part; Lavasanat, Qarah Daq, Shemiranat, Hassan-Abad and the Namak mountains are located in the southern part of the region. Bibi-Shahr-Banoo and Alqadr are located in the southeast and the heights of Qasr-e-Firoozeh are located in the eastern part of the province. In terms of environmental conditions, the climate of Tehran province is extremely diverse. For instance, in the southern quarter the climatic conditions are dry and warm, whereas in the vicinity of the mountains it is semi-humid and cold; the higher region experiences extremely cold and long winters. Mid-July to mid-September are the hottest months of the year when temperatures range between 28 and 30°C; and the coldest months average 1°C in December and January, although at times it might reach −15°C in winter (Fitzgerald et al. 2000). As a matter of fact, climate plays an important role for people settlement and development. Average annual rainfall precipitation is approximately 200 mm, peaking during the winter season. In general, this province has a semi-arid, Steppe climate in the south and an alpine climate in the north (Iran Chamber Society 2001).

3.4 Transportation

The importance of pointing out the issue of transportation is because transport dimensions in Iran, and specifically in Tehran, show marked differences in comparison with European models. Urban growth in Tehran area has a close relationship with its road networks, because the densely used transport system is extremely reliant on cheap oil prices, affecting people's mobility habits and other related matters consistent with urban expansion (Bertaud 2003). As demonstrated in Fig. 3.3, Tehran is a central connection point between the west and east of the country, which again exacerbates the intense internal road developments and prolific external freeways.

Public transportation in Tehran largely consists of taxis, large and small buses, and the subway system, but using personal vehicles has steadily become more pervasive due to cheap oil prices—hence, people have become more dependent on the car. Inconsistencies in the overall public transport system have also played a part in the shift to using personal transport.

More recently, however, initiatives to improve and expand a more comprehensive public transport plan are being implemented. Additionally, recent government policies towards removing oil subsidies, thus pushing up the cost of fuel, have made the development of an improved public transport even more essential.

The Metro Company which operates the subway system is an affiliated division of Tehran municipality and is responsible for all subway extensions. There are four subway lines in Tehran, one of which connects the west suburbs (e.g. Karaj) area to Tehran (Tehran Urban and Suburban Railway Co 2010). In terms of air travel Tehran has four airports, including two main public airports, such as the Mehrabad International Airport, Imam Khomeini International Airport, as well as the military airports at the Ghaleh-Morghi airfield and Doshan-Tapeh airbase.

The metropolis of Tehran benefits from a massive network of highways (around 285 km) and intersections, over-ramps, and 'spaghetti' loops (about 180 km); however, in 2007, around 130 km of highways and 120 km of over-ramps and intersections were under construction (Tehran Municipality 2010). The centre of the city also hosts primary government departments such as ministries and

Fig. 3.3 Transportation network in the Tehran metropolitan area

municipal headquarters (Iran Chamber Society 2001) which adds to the density of the entire transport network.

3.5 Climate

Based on the Köppen climate classification, Tehran is located in a semi-arid, continental climate. In general, its climate is described as mild springs, hot and dry summers, pleasant autumns, and cold winters. This large city has considerable differences in elevation among a variety of districts. Its weather is usually cooler in the mountainous north in comparison with the flat southern part of Tehran. The majority of the annual rainfall occurs from late-autumn to mid-spring. Summer is typically warm and dry with minimal rain, but relative humidity is frequently low and the nights tend to be cool. July is the hottest month of the year with an average minimum temperature of 23°C and average maximum temperatures of 36°C. January is the coldest month with a mean minimum temperature of −4°C and also, a mean maximum temperature of 6°C. Compared with other areas in the country Tehran benefits from a mild climate, but weather conditions can sometimes be surprisingly harsh. Record high temperature can reach 48°C with lows of −20°C.

3.6 Demography

Demographically Tehran is essential for this kind of formal research, because of the significant shift in population from rural to urban living, and the concomitant growth of built-up accommodation. Therefore, a detailed demographic analysis will be presented in Chap. 4. Tehran is the economic centre of Iran and also, about 30% of Iran's public sector employees, and 45% of large industrial companies, are located in the metropolis.

Almost half of this workforce consists of government, public sector employees. In recent years, many modern industries, such as automobile manufacturing, electronic equipment, textiles, armaments, cement, and chemical products have been developed in the vicinity of Tehran (Iran Chamber Society 2001).

3.7 Pollution

Although it might seem unnecessary to deal with Tehran air pollution within the scope of this research, it is an issue which is one of the consequences of LUCC in this area, and is, accordingly, of extreme influence in the city. The city of Tehran suffers from severe air pollution and is frequently covered with smog, which has a debilitating effect on the health of the people. It is estimated that around 27 people die every day through pollution-related diseases. According to local administrators, 3,600 people died in a single month due to the toxicity of the air quality. Some 80% of the city's air pollution is a direct result of carbon emissions from automobile and the rest is due to industrial pollution. In 2007, Iran imposed fuel rations and price rises, but the plan has achieved little success in reducing the pollution levels and the proliferation of personal transport.

The air pollution is primarily due to the following reasons:

- *Geographical*: Tehran is bordered by the massive Alborz mountain range in the north which prevents the flow of the humid winds coming from the Caspian Sea. The UV radiations, combined with existing pollutants, considerably raise the level of the ozone.
- *Economic*: Most Iranian industries are located on the fringe of Tehran. Many old cars are still being used which do not meet standard emission regulations. Additionally, the busiest airport, Mehrabad International Airport, is located in the city.
- *Infrastructure*: Tehran has no well-organised public transport network. Buses and metros are unable to cover every location within the city. Therefore, most people are forced to either hire taxis or buy and use private cars. This has created severe traffic congestion and consequently higher rates of pollution.

3.8 Tehran Spatial Structure

Three main features can illustrate the spatial structures of urban areas:

- the spatial distribution of population in built-up areas;
- land per capita (i.e. the consumption of land per person);
- the concentrated pattern of daily trips within the city (Bertaud 2003).

Latest reports concerning Tehran's spatial structure demonstrate exceptional spatial structure for this city: It has a high density structure combined with a slightly polycentric pattern. The lack of a strong and centralised core, as well as the spatial dispersion of employment, are generally associated with built-up density. Tehran's built-up density, approximately at 146 person/ha, is highly unusual for a polycentric city. This atypical characteristic cannot be necessarily negative, although it suggests it would be necessary to find some comparative solutions that have worked out successfully in other cities with a completely dissimilar spatial structure. The current irregular spatial structure of Tehran which is a high-dense city without a dominant Central Business District (CBD) creates several restrictions. At the central administration level, the alternative of restricting the population of Tehran to 7.6 million people (population in 2002) has been discussed. However, this is not a sustainable option since the government has not been able to control demographic growth (Bertaud 2003).

3.9 Land Consumption Per Person

The area of land consumption per person can be interpreted as the area that a city requires for its expansion. It is typically measured by its inverse, the density per person measured in people per hectare of built-up areas. The density in the built-up areas within the municipal boundary of Tehran is approximately 146 people per hectare. Compared with the built-up density of other cities in the world, this is a rather high density ratio (see Fig. 3.4).

However, compared with other large cities of Asia, Tehran's average density is almost twice as low as current built-up densities in east Asian metropolitans such as Seoul. Hence, this comparison shows that the density of Tehran should not be the main issue (Bertaud 2003). Besides, Tehran has two main types of expansion which are firstly, to spread the city territory, and secondly, expansion of the third dimension of the city: i.e. the construction of high rise and middle range housing neighbourhoods. This has become very common in recent years which afford huge commercial benefits for land developers.

Bertaud (2003) calculated the evolution of Tehran built-up densities between 1891 and 1996. The recent changes will be considered more accurately in the next chapter. But Fig. 3.5 demonstrates an overall change pattern during the mentioned period.

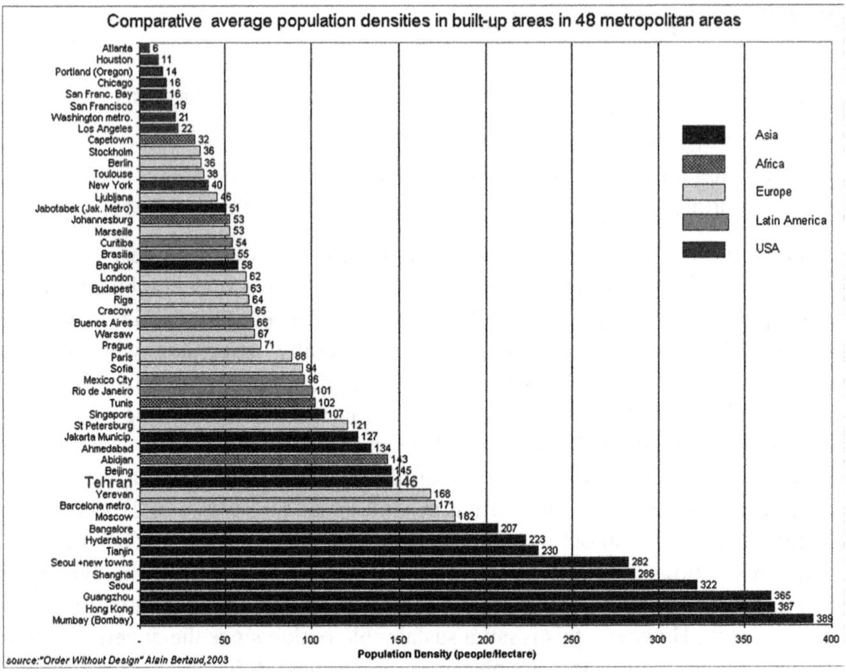

Fig. 3.4 Comparative average population densities in built-up areas in 48 metropolitan areas (Bertaud 2003)

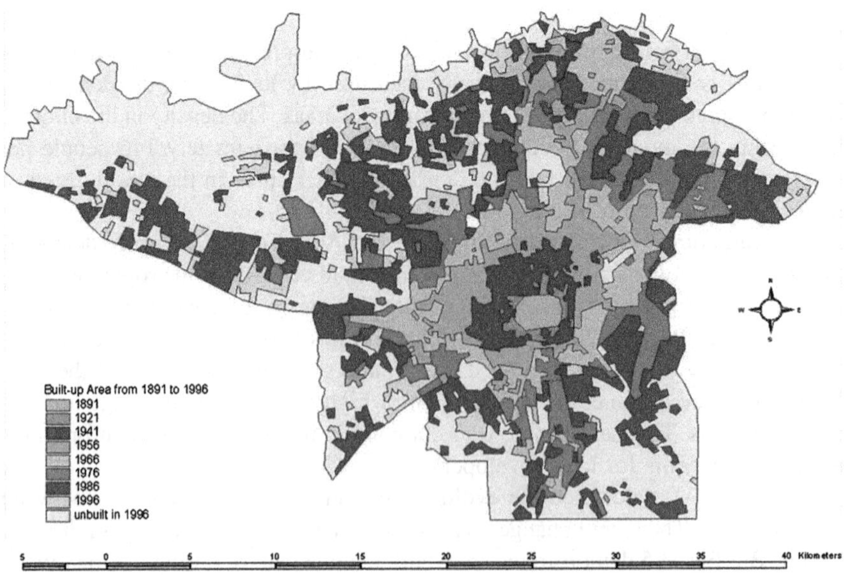

Fig. 3.5 Built-up area changes from 1891 till 1996 (Bertaud 2003)

Figure 3.6 expresses the density of Tehran has been diminishing within the period from 1860 to 2000, stabilising around 1950, and began increasing somewhat between 1970 and 1976, and again stabilising at its present level between 1987 and 1996.

High rise buildings have emerged since 1960 and have multiplied since 1980. Bertaud (2003) noted that

> This increase in building height have resulted in lower density—and not higher densities as would have been expected—because simultaneously as building were getting taller, Tehran's households consumed more floor space per person.

The present mean floor consumption in Tehran is 25.5 m^2 per person, similar or higher than the consumption of a number of cities of Europe. The increase in floor consumption per person is higher than the increase in floor space density. The construction of high residential building tends to reduce housing prices and, therefore, increase floor consumption per household (Bertaud 2003).

3.9.1 Spatial Distribution of Population

The population distribution in Tehran does not match typical patterns with a decreasing rate toward city borders (Bertaud 2003). Figure 3.7 demonstrates that the spatial pattern of Tehran metropolis has the capacity to fill up unoccupied and open spaces.

In the case of the majority of cities, the central business district (i.e. CBD) is located within the heart of the city's environs. The intensity of business and retail

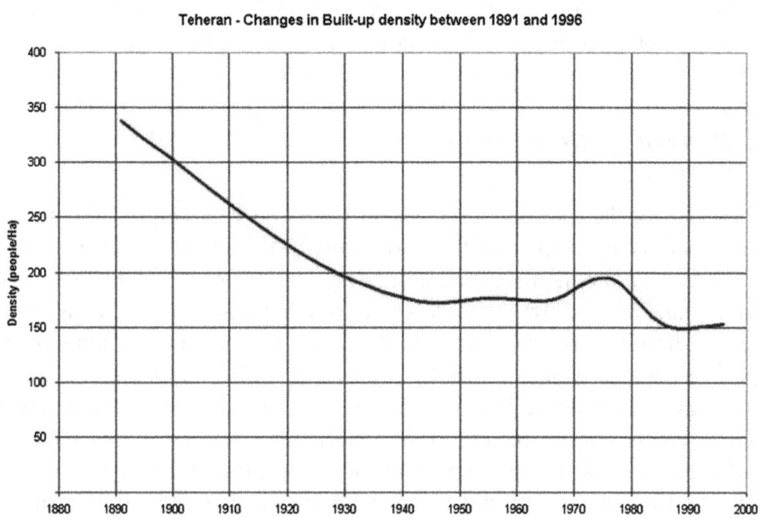

Fig. 3.6 Changes in built-up area density between 1891 and 1996 (Bertaud 2003)

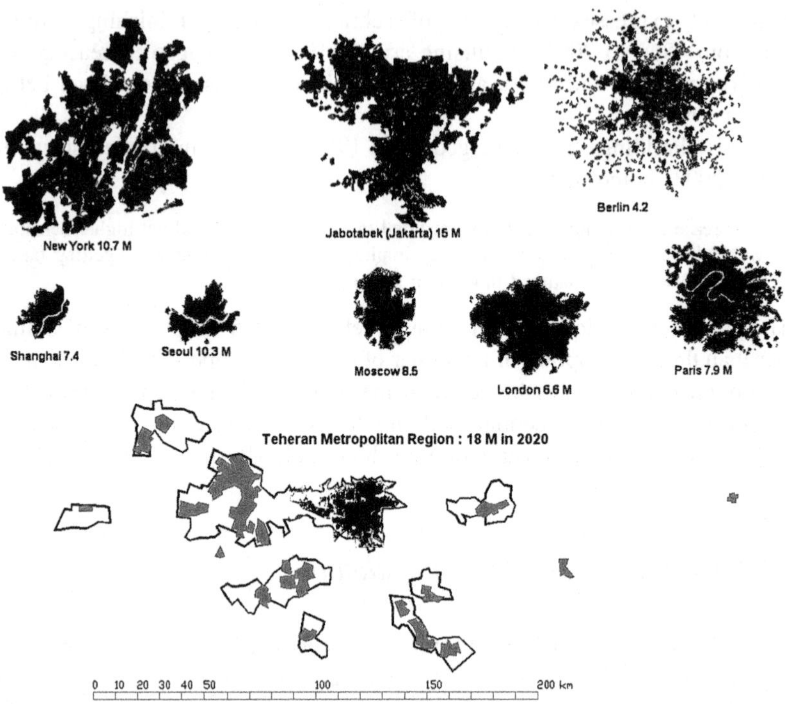

Fig. 3.7 Tehran metropolitan area in comparison with other metropolises (Bertaud 2003)

use which distinguish the CBD is absent from central Tehran and also, demonstrates that the fairly low-density CBD area is more likely due to abandoned lands and buildings or large institutional holdings (Bertaud 2003).

3.9.2 Pattern of Daily Trips

An awareness of the transportation system would be very helpful in realising the residents' behaviour in terms of their proximity to the city centre or their working areas. Whereas at the moment the transport system costs are still being subsidised and do not take a big share of people's lives in comparison with foreign countries, people can, however, purchase houses far outside the city and commute by public transport or personal vehicles. On the other hand, whilst the public transportation system is affordable, traffic jams still plague regular daily trips, which makes it a difficult dilemma for inhabitants to deal with this issue, i.e. living close to the centre of the city and paying extra rent, or living outside the main core, beside freeways and highways with more affordable housings and the option of buying a car.

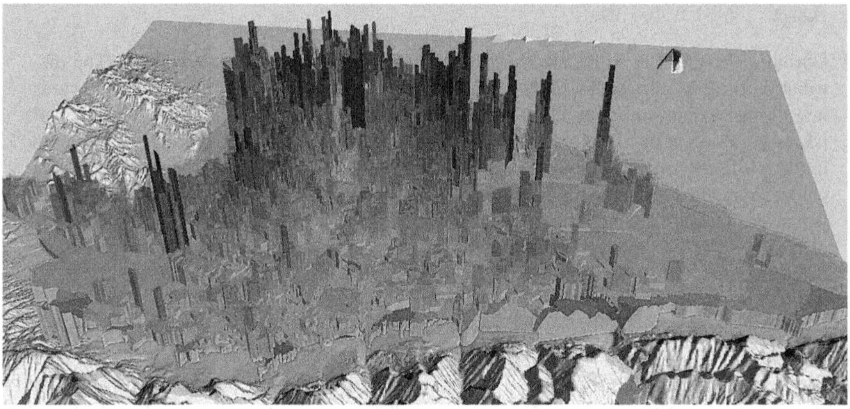

Fig. 3.8 Spatial distribution of population density in the study area

Destination and origin matrices for vehicular job commuting and shopping trips explain that districts 6 and 12 comprise the closest districts to a CBD in Tehran. Also, these two districts serve only 27% of all shopping expeditions, as well as 30% of all job commuting journeys. If jobs were equivalently distributed among all 23 districts assessed, each district might obtain 5% of all journeys. It has to be noticed that the matrix shows only vehicular trips and not all trips. Given the high population density of Tehran, an enormous number of shopping trips take place on foot to neighbourhood shops and, therefore, the 27% figure of vehicular shopping trips to the CBD is possibly an overestimate of all the shopping trips undertaken. A 3D visualisation of the study area, which represents the spatial distribution and its density, is shown in Fig. 3.8.

3.10 Ancillary Information

A number of urban landscape regulations have been issued since the Islamic revolution of 1979; in addition, new urban land development policies have been initiated in Iran, and are regularly updated. These policies have had some major purposes (e.g. sustainable urban development). Besides, Iran has experienced a remarkable growth in urban population in recent years, proving that adequate land supply has become an important concern for urban land development policies. The effects of centralisation can be seen at different levels of the land development process (e.g. site selection, design, agreement, the financing process and implementation). Furthermore, all project agreements have to be permitted by the Ministry of Housing and Urban Development, where large scale constructions have to have the approval of this organisation (Azizi 1998).

Azizi (1998) noted that;

> During 1979–85, Iran confronted several problems in urban land development. The distribution of mass raw land by the ULO (Urban Land Organization) contributed towards more urban expansion without adequate infrastructure or consideration of environmental issues, resulting in the development of shanty towns. The trend of rapid population growth in the last few decades suggests that population growth will be the dominant demographic characteristic that will affect housing demand in the future. A positive response to both the extent of housing demand and the need for sustainable development is dependent on the provision of land and infrastructure, with associated environmental protection and enhancement.

According to the 2006 demography reports, Tehran, as the capital of Iran and Tehran Province, has a population of 8,429,807; it is also one of the largest cities in Western Asia, and according to the city mayor's website is the 20th largest city in the world in 2007. As well as being the centre of most Iranian industrial, chemical and core manufacturing, Tehran is also the most important centre for the making and selling of carpets and furniture; therefore, the city manages a massive range of businesses in Iran and nearby countries, for internal consumption and export. During the twentieth century, Tehran was subject to mass-migration of people from other parts of Iran, which included a variety of cultures, peoples and religions. According to statistics from the website of "City mayors" in 2009, Tehran was the 33rd most expensive city in the world. Moreover, it globally ranks 16th in terms of city population, 56th in terms of the size of GDP, and 29th in terms of metropolitan population (City Mayors Statistics 2010). Resulting high inflation rate in the country (which approximately exceeds 17%) can cause to raise its rank to higher levels and distract this list continuously.

Tehran province is known as the financial centre and the richest province of Iran, which contributes around 29% of the country's GDP. Furthermore, it resides approximately 18% of the country's population. Tehran Province is the most industrialised province in Iran; 86.5% of its population houses in urban areas and 13.5 % of its population resides in rural areas. It is considerable that lately some policies have been taken into account in order to supply some motives to dispatch people to the other nearby cities and evacuate Tehran's population due to the risks of probable earthquakes.

3.11 Summary

This chapter has described the study area, where is Tehran, and its situation (Sect. 3.2). A brief explanation about the geography of the selected study area and other affiliated matters to this research such as transportation, climate, demography, and pollution were explained (Sects. 3.3–3.7). The spatial structure of Tehran and the spatial distribution of inhabitants were also discussed (Sects. 3.8 and 3.9).

References

Azizi MM (1998) Evaluation of urban land supply policy in Iran. Int J Urban Reg Res 22(1):94–105

Bertaud A (2003) Tehran spatial structure: constraints and opportunities for future development. Ministry of Housing and Urban Development, Tehran

City Mayors Statistics (2010) The largest cities in the world and their mayors. http://www.citymayors.com/statistics/largest-cities-mayors-1.html

Fardi GRA (2010) Current situation of air pollution in Tehran with emphasis on district 12. Institute for global environmental strategies (IGES), First meeting of the kitakyushu initiative network, Japan

Fitzgerald P, Hickey A, Jenkins MA, Holland T, Andy (2000) Tehran province geography. http://iguide.travel/Tehran_Province/Geography

Iran Chamber Society (2001) Cities of Iran, provinces of Iran. http://www.iranchamber.com/provinces/01_tehran/01_tehran.php

Shahshahani S (2003) Tehran: paradox city. IIAS Newsletter N31:15–16

Tehran Urban & Suburban Railway Co (2010) http://www.tehranmetro.com/Default.aspx

Tehran Municipality (2010) Spatial data infrastructure organization center. http://sdi.tehran.ir/

Chapter 4
Data Preparation and Processing

4.1 Introduction

This chapter first comprises an overview about the available data, data characteristics, the source of data and the preparation procedure with explanations. Then, the trend of change will be analysed by means of temporal mapping through Landsat images at different times. The amount of urban sprawl will be measured and predicted for forthcoming years.

4.2 Data Acquisition and Data Collection

In this research, various sorts of data such as multi-spectral and temporal satellite images, a set of environmental, terrestrial attributes, and socioeconomic data were gathered. A temporal coverage of Landsat imagery from different sensors was collected. This temporal coverage includes satellite images of MSS, TM, and ETM^{+} within the past 27 years. These satellite images were acquired through the Earth Science Data Interface of the Global Land Cover Facility, and also, the Earth Resources Observation & Science Centre (EROS) of US Geological Survey. A set of high resolution aerial photos of the study area were gathered in order to check ground control points. These satellite images were employed to extract land use and land cover maps. Also, the other collected data included demographic details of Tehran's metropolitan area, extracted through the last accomplished available demography statistics. This sociological information was downloaded from the webpage of the Iranian statistic centre. Additionally, a geodatabase of environmental and urban features such as topography, hydrology, building blocks, transport network, farming land and prepared land use maps was gathered through several sources (e.g. Tehran GIS centre and other affiliated organisations). These data were stored in different scales. A brief description of the utilised data is demonstrated in Table 4.1.

J. Jokar Arsanjani, *Dynamic Land-Use/Cover Change Simulation: Geosimulation and Multi Agent-Based Modelling*, Springer Theses,
DOI: 10.1007/978-3-642-23705-8_4,

Table 4.1 A description of the utilized data in detail

Type of dataset	Resolution/scale	Source
Landsat images (MSS, TM, ETM+)	30 m	EROS
Geodatabase of existing features (e.g. transport network, streams, airports, administrative divisions, etc.)	1:25,000 1:50,000	Statistical centre of Iran, National Cartographic Centre (NCC), Iranian Space Agency (ISA)
Land use map	1:50,000	Tehran GIS centre
Digital elevation model	30 m	Tehran GIS centre
Demography data	1:50,000	Statistical centre of Iran

4.3 Data Processing

The prepared geodatabase of urban features and environmental elements that were collected through multiple sources had to be matched in terms of a geodetic reference system, scale and file format. ESRI Shapefile was chosen as the base file format and all the data were converted to this format. Topographic data, after preparation and missed data corrections, were converted to a digital elevation model (DEM). A set of topographic factors such as slope, aspect, and hillshade was produced, and all the gathered data were imported into the geodatabase.

Remotely sensed imagery is a generally recognised crucial source for land use change monitoring. The aforementioned satellite images were collected and stored on a hard drive in order to do signal processing and remote sensing analysis to achieve land use/cover maps. Therefore, after a preview of on-hand images and the removal of cloudy ones, a combination of images were produced for temporal analysis. A regular set of 10-year stage images of 1986, 1996, 2006 was chosen, and the gathered demography and socioeconomic data were converted to spatially explicit data with the aim of compatibility with other geodatabase datasets. The gathered data were controlled in terms of quality and certainty (e.g. data georeferencing). However, it should be noted that in this section this procedure is simply touched upon.

4.4 Temporal Land Use Map Extraction Through Remote Sensing

The prepared land use maps of 1986, 1996, and 2006 were obtained through the Tehran GIS centre. After a comparison with the actual situation of Landsat satellite images, some misclassifications were found in the maps. Hence, it was vital to update these land use maps with the Landsat images and also the aerial photos to achieve more accurate land use maps. These maps were classified into six categories such as open land, agricultural land, water bodies, industrial area, residential area and public parks.

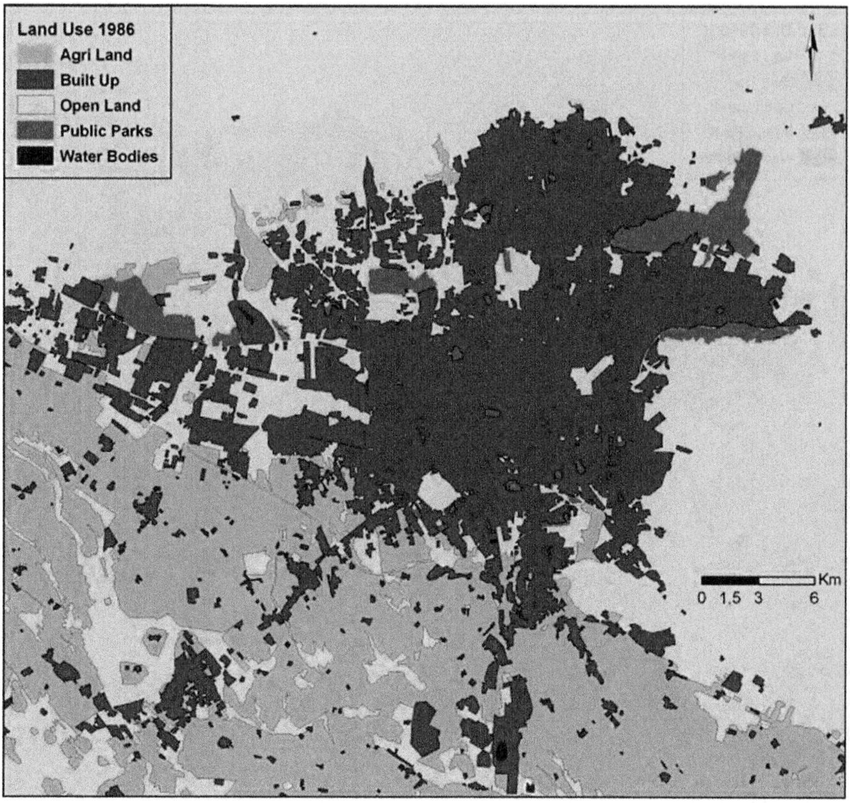

Fig. 4.1 Final extracted land use map of 1986

The finalised land use maps were eventually produced by overlaying remote sensing data, prepared land use maps, implementing various remote sensing techniques. These maps are the source maps of this research which are shown in Figs. 4.1, 4.2 and 4.3. The accuracy assessment process was done through Kappa index calculation. The accuracy of these maps was significant, since they are intended to be utilised as the base input files for simulation. Moreover, the six classes mentioned above were summarized into five categories by combining residential and industrial areas into the built-up class. This process avoided more possible complications arising from the discovery of industrialised zones, and reduced the computation process.

4.5 Temporal Mapping and Changes Visualisation

The prepared land use maps are shown in Figs. 4.1, 4.2, and 4.3. These maps demonstrate the land use pattern of the study area. This temporal mapping onstrates that Tehran is a centralised city with outward expansion. These maps

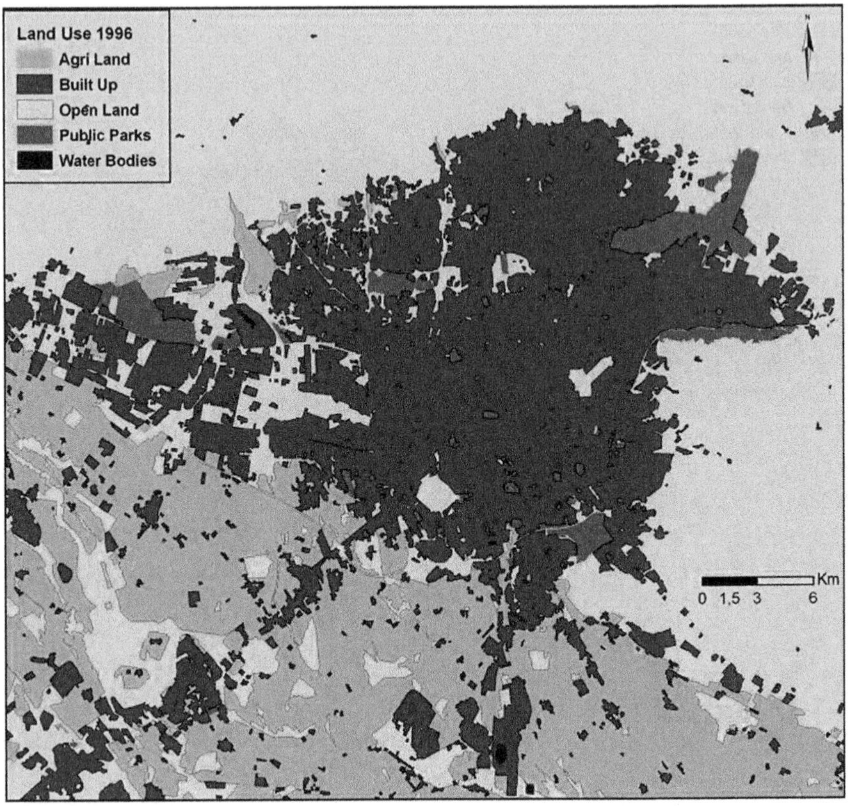

Fig. 4.2 Final extracted land use map of 1996

show that the changes have taken place mainly in north-west towards the south-eastern part of the border (anti-clockwise). In fact, high mountains in the northern and eastern parts of the city do not allow for more expansion; however, some expansion is still taking place.

Finally, the prepared land use maps of 1986, 1996, and 2006 were stored in the created geodatabase.

4.6 Evaluation of Change Trends

This temporal land cover mapping allows us to track the quantity and location of changes. The area of each land class was retrieved and imported to Table 4.2. The quantity of each land category in different years is shown in this table. Obviously, the trend towards the expansion of built-up areas over agricultural lands and open lands has been closely monitored; however, a nominal attempt to extend the

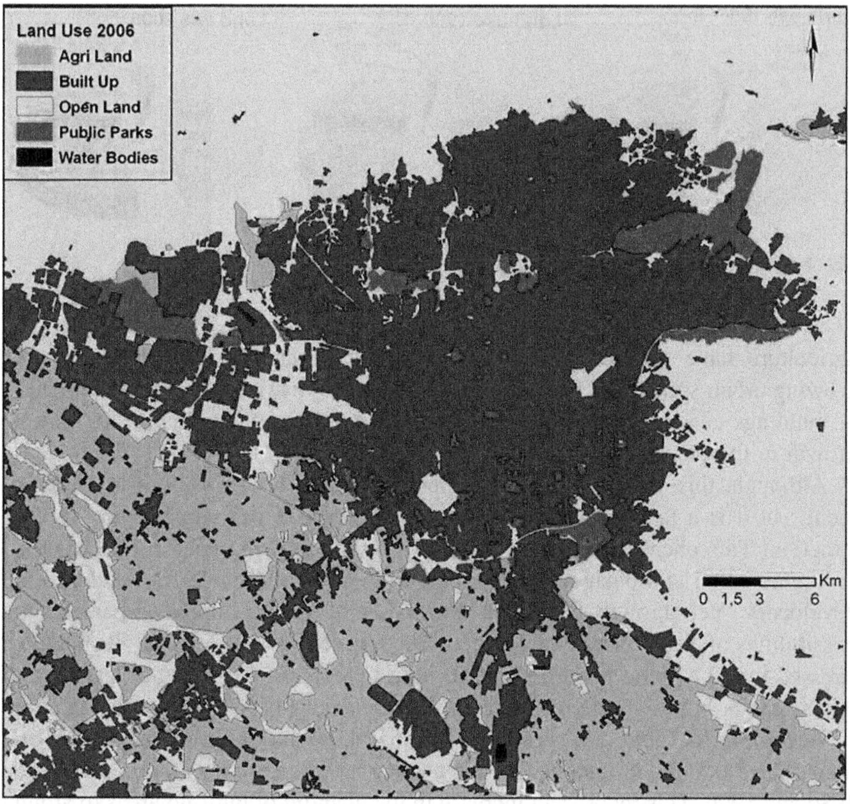

Fig. 4.3 Final extracted land use map of 2006

number of public parks and green leisure areas has been pursued by the municipality.

The results in Table 4.2 and Fig. 4.4 show that the percentage of built-up areas has been accelerating from 24 to 28%, up to 32%. This trend has resulted in the reduction of the farming area and open lands at a rate of 24 and 50%; 22 and 47%; and 21 and 44%, respectively. Consequently the mass of open lands and

Table 4.2 Land use type portions during the past 20 years

Land use type	Area 1986 (ha)	Area 1996 (ha)	Area 2006 (ha)
Agricultural land	45,174	41,172	40,005
Built-up	44,772	53,056	59,095
Open land	92,276	87,305	81,741
Public park	4,076	4,744	5,394
Water bodies	102	123	165

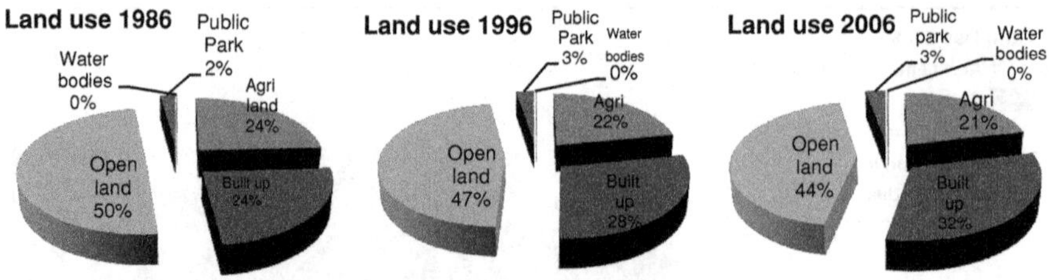

Fig. 4.4 Measurement of land use change between 1986 and 2006

agriculture have suffered on account of urban development. Despite this fast growing urban sprawl the fact remains that in some districts of Tehran, the height of buildings generally is not particularly high; however, the potential for vertical growth in the metropolis remains a real possibility.

Although, this study is not an attempt to measure high-rise built-up development, but it is a fact, nonetheless, that the cityscape is becoming higher in construction. This phenomenon is being driven by the growing need for more urban accommodation and business outlets, which is more affordable for residents and developers. Yet through this third dimension of vertical urban expansion the possibilities of further land change occurrence are minimalised. Again, it should be stressed that this trend towards vertical built-up expansion is not the intention of this research, but might be of interest to urban managers for future consideration.

According to Table 4.3, significantly about 4,000 ha of agricultural fields and more than 5,000 ha of open lands were taken over for urban development between 1986 and 1996. This records an increase in the quantity of built-up areas to around 8,300 ha in total. During this 10-year period, between 1996 and 2006, nearly 6,000 ha of agricultural fields and open land areas were converted to built-up development. This extensive city sprawl is one of the consequences of large migrations towards this metropolis, to meet the needs of the massive influx of people. Additionally, during this 20-year period (1986–2006), approximately 10,000 ha of open land area, along with 5,000 ha of agricultural fields, were converted into built-up areas. This represents an 8% increase in urban built-up environment at the expense of natural land, revealing a growth in urban development from 24% in 1986 to 32% in 2006.

Table 4.3 Quantification of changes between 1986 and 2006 in terms of hectares

	1986	1996	2006	1986–1996	1996–2006	1986–2006
Agricultural land	45,174	41,172	40,005	−4,002	−1,167	−5,169
Built-up	44,772	53,056	59,095	8,284	6,039	14,323
Open land	9,2276	87,305	81,741	−4,971	−5,564	−10,535
Public park	4,076	4,744	5,394	668	650	1,318
Water bodies	102	123	165	21	42	63

4.7 Measuring Change and Sprawl

The acquisition and conversion of farming land and open land on the fringe of cities into built-up areas is, by definition, urban sprawl. It is the aim of this thesis to measure the extent of this phenomenon.

Therefore, in order to predict the extent of this change in these fringe areas, previous occurred and anticipated values of change need to be taken into account. In academic terms and according to Torrens and Alberti (2000).

> Sprawl is a highly contentious issue—neither its determinants nor its characteristics are fully understood.

In recent years, researchers have performed conceptual investigations of the sprawl phenomenon, such as its characteristics, causes, and potential controls. A variety of methods to provide a better understanding of urban growth have been discovered by respective experts to measure sprawl. Theoretical discussions on the subject of sprawl have created a wealth of discussion around the matter. Urban sprawl is a practical, real world issue of considerable concern. Therefore, the sense of urgency that prevails with regard to the sprawl problem, creates an extreme need for more theoretical discussion to improve current methodologies (Torrens and Alberti 2000).

According to Table 4.2, within 20 years, around 14,000 ha of built-up area have been constructed by destroying roughly 50,000 ha of farming area, as well as 11,000 ha of open lands, which could have been used to expand green spaces or farming development.

Although urban expansion brought expectations of more public parks and green areas, no such increase of parkland has been observed.

An increase in the number of parks and green spaces for the Tehran metropolis, to especially combat air pollution, would seem of prior importance, thus urban planners need to take this issue into account. In fact, air pollution accounts for a number of deaths per year. From a landscape planner's aspect, a balance between urban development and the creation of green space should be a priority. However, it seems the municipality of this metropolis was not able to create new green spaces within the second period (i.e. 1996–2006).

4.8 Socio-Demographic Changes

Tehran has been confronted with various socio-demographic changes within recent decades. These changes have been caused mainly by three considerable conditions: first, change in governmental policy over family structure and childbirth numbers following the 1979 revolution; second, the population lost during the Iran-Iraq war, which encouraged people to support this loss; and thirdly, the mass of migration from other cities to Tehran to find work, to study and secure a better

Table 4.4 Tehran city, province and study area population details

	1966	1976	1986	1996	2006
Tehran city population	2,980,041	4,530,223	6,042,584	6,758,845	7,797,520
Study area population	2,981,047	4,580,515	6,257,713	7,024,295	8,154,691
Tehran province population	3,455,537	5,313,143	8,095,124	1,0343,965	13,413,348
Country population	25,788,722	33,708,744	49,445,010	60,055,488	70,472,846
Tehran city share of province (%)	86.20	85.30	74.60	65.30	58.10
Tehran province share of country (%)	13.40	15.80	16.40	17.20	19.00

standard of living. Altogether, these conditions have created significant urban pressures within Tehran to accommodate these socio-demographic changes.

According to Table 4.4, there are some significant conclusions which can be deduced as follows:

1. The population of Tehran city has been growing exponentially since 1966.
2. The population of Tehran province, within the country as a whole, has been increasing, proven by the mass migration towards this province from other provinces.
3. The population share of Tehran city, as of Tehran province, has been decreasing significantly, which means the majority of migrants tend to settle in suburban areas and, moreover, nearby cities have been developing even more quickly than Tehran (see Fig. 4.5).
4. The overview determines that nearby cities are housing a huge and growing number of provincial inhabitants, and their proximity to the metropolitan core is becoming increasingly less, i.e. this trend can cause a continuous homogeneous metropolis with many complications.

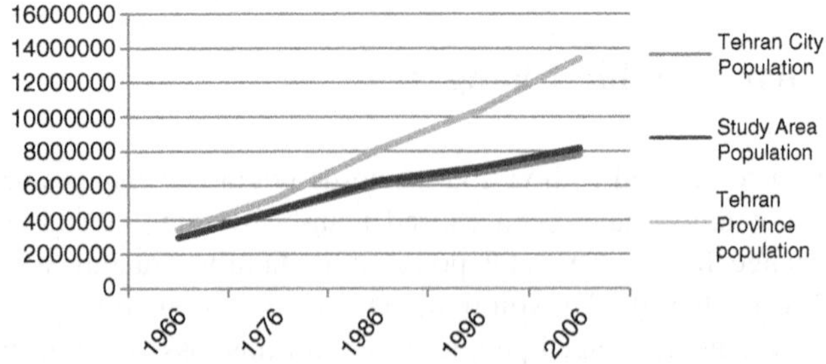

Fig. 4.5 Population growth in Tehran city in comparison with the metropolitan area

Table 4.5 Population density per area unit

Year	Residents population (Person)	Occupied area (ha)	Construction per capita (Person/ha)
1986	6,257,713	44,772	139.7
1996	7,024,295	53,056	132.4
2006	8,154,691	59,095	138.0

4.9 Measuring Per Capita Construction

Construction per capita is an often reported and commonly compared statistic for measuring the number of domestic residents in a city. Population density is the measurement of the number per unit area. It is commonly represented as people per hectare/square kilometre, which is consequent simply by dividing the total area population (i.e. those who have settled in a region) by region area. This calculation implements a straightforward mathematical method which is given in Eq. 4.1.

$$Construction\ per\ capita\ (Person/ha) = \frac{\text{Residents population (Person)}}{\text{Builtup Area (ha)}} \tag{4.1}$$

Equation 4.1 was executed to obtain this index for the past. The mean value of the three past periods was used to predict this index for forthcoming periods (see Table 4.5). Thus, by means of this index, as well as predicted population, the quantity of developed lands for the future can be determined, which is shown in Sect. 4.10.

4.10 Estimation of Change Demand

The quantity of change demand can be calculated through two possible ways: by retrieving via the Markov chain model, or through the extrapolation of statistical demographic data.

- *Markov chain model*: This model is able to predict the next status of change according to the previous status; therefore, a transition area matrix is produced by this model indicating the amount of change between existing categories. Table 4.6 presents the anticipated amount of change for the forthcoming period, which is in this example 2016. This matrix indicates that approximately 1,605 ha of agricultural land will be replaced by built-up areas by 2016. The same table for 2026 has been produced.
- *Statistical extrapolation*: The other possible method to predict land change demand is to calculate the index of construction per capita for the future. Then, the average value of this index can be calculated to input the equation, to incorporate the predicted population for 2016, 2026 as provided by the Iranian statistic centre. The index was calculated in Sect. 4.9 for the previous time points (i.e. 1986, 1996, and 2006). Hence, it is feasible to predict the anticipated amount of change, the

Table 4.6 Transition areas matrix of the Markov model for 2016 in terms of hectare

	Agricultural lands	Built-up	Open lands	Public parks	Water bodies
Agricultural lands	37,424	1,605	871	69	9
Built-up	214	58,326	324	206	2
Open lands	1,175	4,702	75,355	448	26
Public parks	10	47	35	5,290	7
Water bodies	4	2	0	1	158

Table 4.7 Statistical extrapolation of construction per capita index for 2016 and 2026

Year	Residents population (Person)	Occupied area (ha)	Construction per capita (Person/ha)
1986	6,257,713	44,772	139.77
1996	7,024,295	53,056	132.39
2006	8,154,691	59,095	137.99
2016	9,940,964	72,711	136.72
2026	11,553,771	84,508	136.72

assumption being to input the average construction per capita index for the next steps (i.e. 2016, 2026). Therefore, population was statistically predicted to obtain the expected amount of change. This means that change was quantified by this technique. In Table 4.7, the estimated amount of resident population in the study area, and construction per capita index are shown. In fact, the demand of change for the built-up class alone can be projected through this method, whereas the area of the other existing classes could not be estimated in this way. Thus, the Markov model is able to predict quantity of change for each particular land type.

4.11 Summary

In this chapter, the process of preparing the utilised data has been explained and also, a temporal mapping of land use change in the study area represented. The quantity of change has been analysed statistically and, therefore, the urban sprawl accordingly measured. In addition, socio-economic changes within the selected time periods have been taken into consideration.

As a final point, the forthcoming changes have been estimated through two different scenarios to be employed in the change allocation process.

Reference

Torrens PM, Alberti M (2000) Measuring Sprawl, CASA Working Papers (27). Centre for Advanced Spatial Analysis (UCL), London, UK

Chapter 5
Implementation of Traditional Techniques

5.1 Introduction

The inherent aim here is to implement the traditional and common methodologies which have been employed to illustrate land use change, and thereafter to simulate its forthcoming status. In this chapter, the cellular automata model, the Markov chain model, the CA-Markov model and the logistic regression model will be designed and executed. Each single model will be evaluated to verify its outcomes. This will allow us to validate their results and acquire enough assurance of their performance. Thus, verified models will be chosen in order to integrate in the ABM.

5.2 Selected Techniques for Implementation

In this part of the chapter, it is intended to review and also execute preferable and useful methodologies such as cellular automata, Markov chain, cellular automata Markov, and logistic regression models. The outcomes of these models will be evaluated and the different results will enable us to compare them to each other. The strategy of creating different results by means of different techniques will enable this research to represent various methods upon a specific area. Therefore, a general flowchart for this section can be presented as in Fig. 5.1.

In Sect. 5.3, we will describe the theoretical background of the aforementioned models, as well as their implementation, starting with the cellular automata model.

J. Jokar Arsanjani, *Dynamic Land-Use/Cover Change Simulation: Geosimulation and Multi Agent-Based Modelling*, Springer Theses,
DOI: 10.1007/978-3-642-23705-8_5,

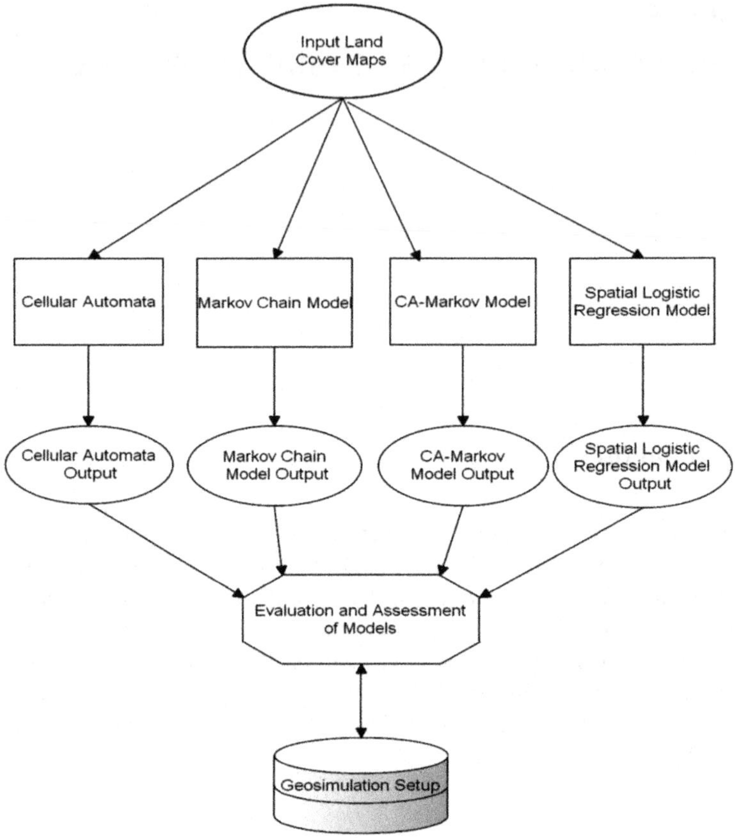

Fig. 5.1 Flowchart of the general strategy for the implementation of the models

5.3 Cellular Automata Model Scenario

In recent decades, investigations for developing geographical cellular automata in order to simulate complex systems have been raised. Cellular automata have been employed to simulate wildfire propagation (Goodchild et al. 1996), population dynamics (Couclelis 1985), and land use change (Batty and Xie 1994; White and Engelen 1993).

The cellular automata model is known as CA which is a dynamic model originally conceived by Ulam and Von Neumann in the 1940s to afford a formal framework for investigating the behaviour of complex systems (Moreno et al. 2009). CA is also the main framework of agent-based modelling scenarios. Land use changes simulation using CA is a complicated process, whereas various spatial variables and factors have to be employed (Li 2008). A critical matter in CA modelling is defining appropriate transition rules based on training data. In fact, these transition rules conduct this model. Linear boundaries have been used to

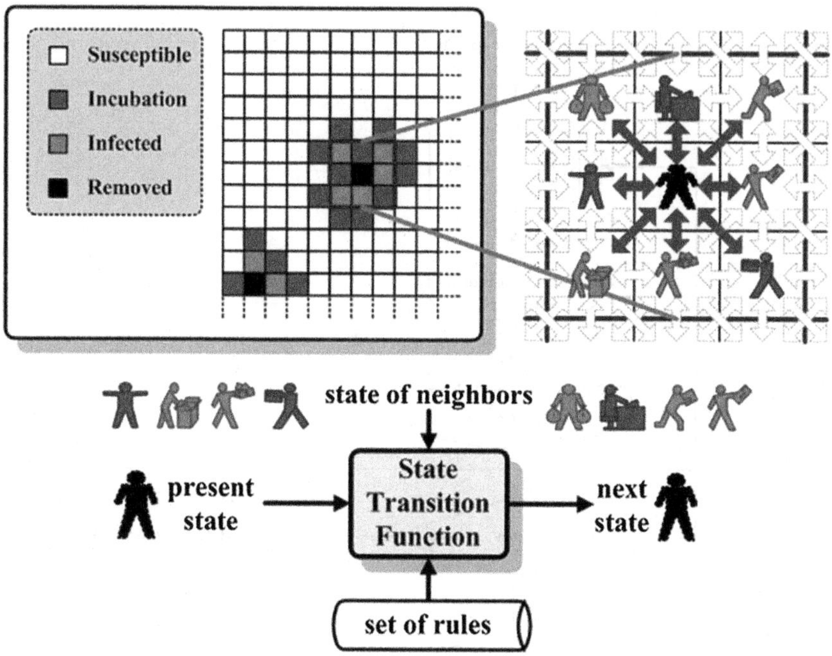

Fig. 5.2 Cellular Automata, state transition rules and the Moore neighbourhood notion (Huang et al. 2004)

define the rules (see Fig. 5.2). However, land use dynamics or changes, and many other geographical phenomena, are vastly complex and require nonlinear boundaries for the rules definition (Moreno et al. 2009). Figure 5.3 demonstrates the flowchart of implementing the CA Model.

5.3.1 CA Transition Rules

Land use changes on the fringe of cities (i.e. urban sprawl) is the consequence of both internal and external forces; the internal impact means an area tends to continue its development if it has begun to develop from a rural to an urban status, particularly if this natural tendency is supported by development from within the neighbourhood. The external impact means factors such as the geographical conditions of the area, socio-economic circumstances and institutional controls, also impact on the process of development. Physical constraints (e.g. water bodies and steep terrain, etc.) restrict or slow down the development of urban areas (Fig. 5.4).

Socio-economic factors, such as land availability and demands on available lands, accessibility to nodes of employment, accessibility to public services and facilities, such as schools, shops, public transport, and contiguity to existing urban areas also play key roles in urban development; therefore, they are able to define

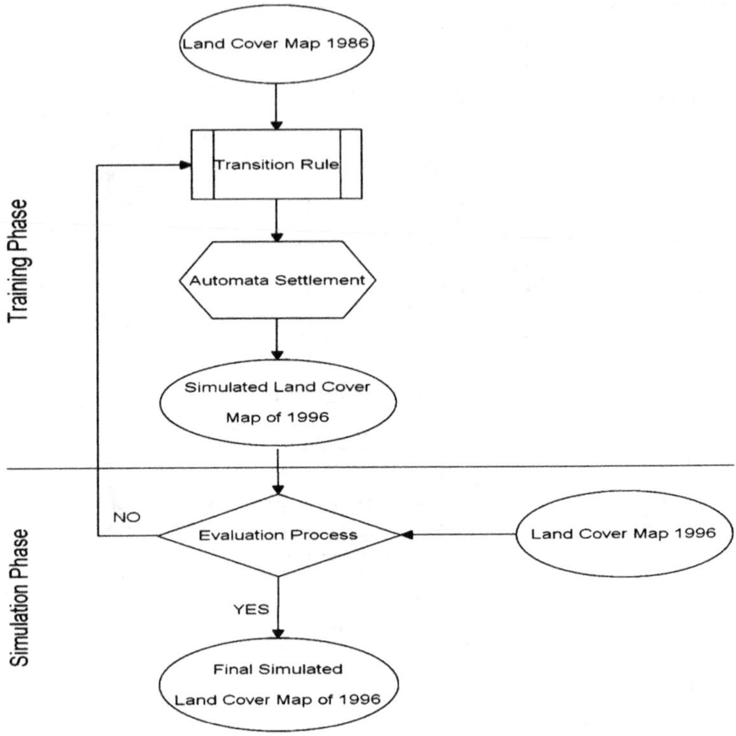

Fig. 5.3 Flowchart of the implementation procedure of the CA scenario

appropriate conditions (Liu 2008). The transition rules are the major inputs in a CA model. Basically, the aforementioned rules have been defined in linear forms, using methods such as multi-criteria evaluation (MCE) (Yang et al. 2008). Transition rules can be defined through a filter file at a variety of kernel sizes, and various decision rules can make that CA model completely different from other existing CA models. Whereas these simulated maps are on hand, a training phase can be utilised by means of these preliminary results and the map of reality. This training phase helps to realise the appropriate kernel size and transition rules.

5.3.2 Training Process and Calibration of the CA Model

The training process consists of choosing a certain time step for the simulation through the CA model. Different transition rules and neighbourhood distances can result in various outcomes; thus, a preliminary evaluation of the obtained results was carried out to pick the optimum settings. The optimum settings will lead us to implement this model by coordination of time step and results. Accordingly, after

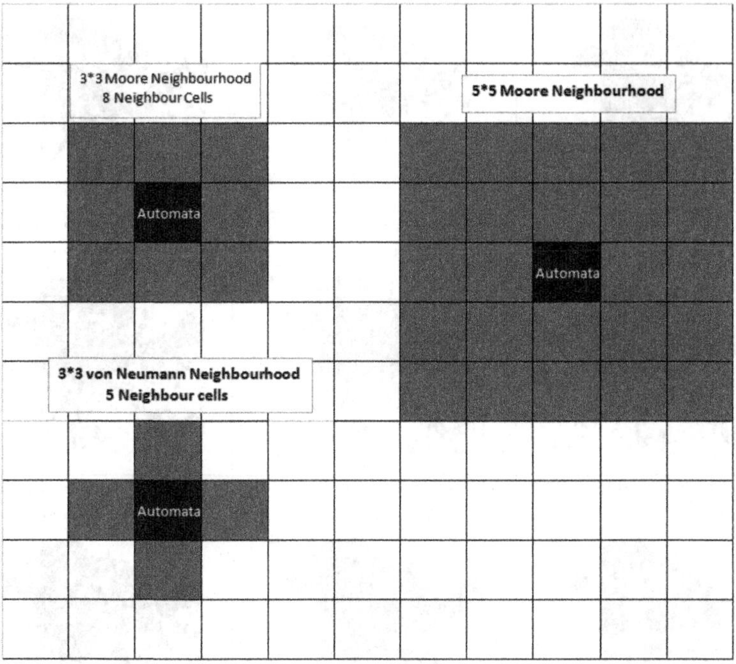

Fig. 5.4 Schematic explanation of automata and different neighbourhood layouts

implementing the training phase and retrieving calibrated factors, a simulated map of development of forthcoming years was prepared.

The other key issue to implement a CA model is to estimate an appropriate iteration number. This enables users to stop the modelling process at accurate times. Therefore, a training process is applied to the model in order to control the predefined transition rules. This helps to stop our model at a certain time and reach a certain amount of change, to better estimate the locations of changes. A code was written in the Python environment and imported into the ArcGIS Toolbox. This script has the typical characteristics of a CA model. The code comprises all cellular automata components, i.e. neighbourhood size and transition rules. This CA code performs according to a predefined iteration number, and it stops at a certain time. At each time step, a filter is applied to the entire image then the output image is reclassified according to the reclassification file, and the produced output image is then used as an input for the next iteration. The process goes on until the predefined iteration number is reached.

Results from different settings can be evaluated and compared with the actual map. This model was implemented to the land use maps of 1986 and 1996 to achieve the simulated maps of 1996 and 2006, respectively. The simulated maps of built-up areas at different iteration numbers of 1986, 1996 and 2006 are shown through Figs. 5.5 and 5.6.

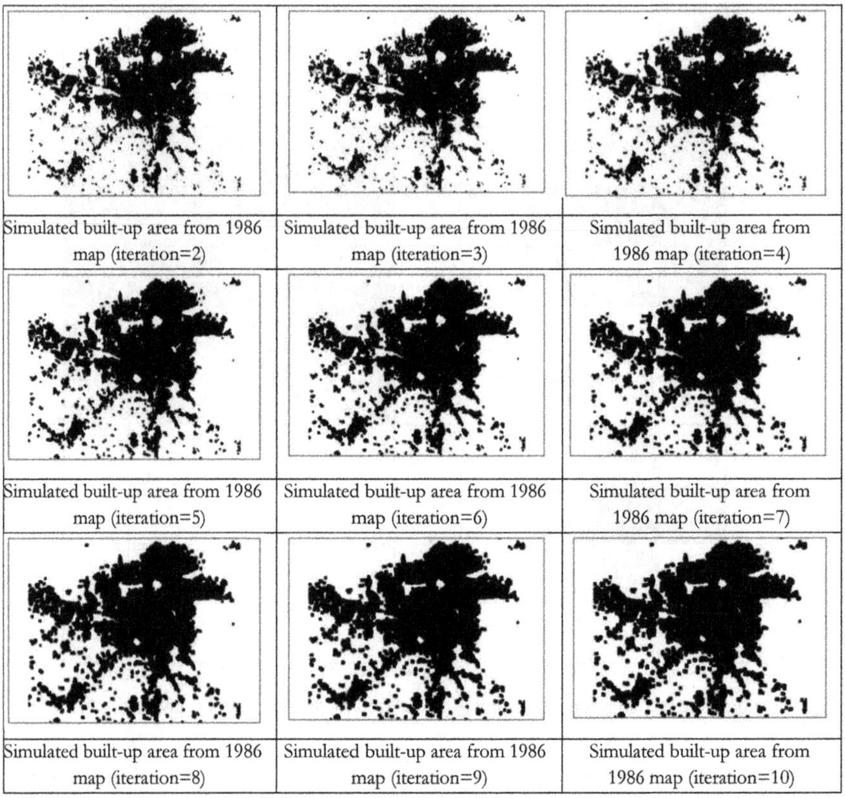

Simulated built-up area from 1986 map (iteration=2)
Simulated built-up area from 1986 map (iteration=3)
Simulated built-up area from 1986 map (iteration=4)
Simulated built-up area from 1986 map (iteration=5)
Simulated built-up area from 1986 map (iteration=6)
Simulated built-up area from 1986 map (iteration=7)
Simulated built-up area from 1986 map (iteration=8)
Simulated built-up area from 1986 map (iteration=9)
Simulated built-up area from 1986 map (iteration=10)

Fig. 5.5 Simulated maps of built-up areas at different iteration numbers from 1986 to 1996

The simulated maps of 1996 and 2006 were compared with the maps of reality of 1996 and 2006; therefore, the optimum transition rules and settings can be determined. In Table 5.1, the number of iterations and resulted ROC values are cross compared to pick the optimum iteration number. The determined transition rules will be chosen as the optimum designed CA model. This model will be implemented on the map of 2006 in order to simulate built-up map of 2016.

Table 5.1 shows that the maximum ROC value yielded at iteration number nine, therefore, this amount of iteration and associated transition rules were employed for the prediction process. The model validation process and resulted map will be presented in Chap. 7 (see Sect. 7.5.1).

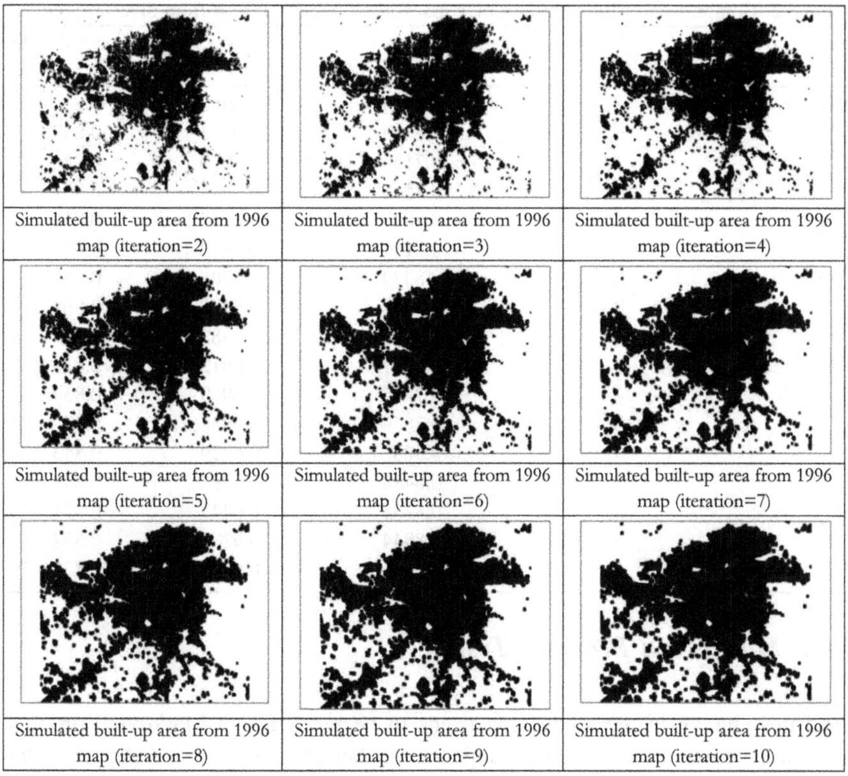

Fig. 5.6 Simulated maps of built-up areas at different iteration numbers from 1996 to 2006

5.4 The Markov Chain Model Scenario

Markov chain theory is a stochastic process theory that describes how likely one state is to change to another state. The Markov chain has a key-descriptive tool which is its transition probability matrix (TPM). Markov chain theory has been used generally to study water resource systems and simulate precipitation sequences, particularly to describe and predict lithological transition, plant succession, and land utilisation change (Li et al. 1999).

Stochastic processes generate sequences of random variables$\{X_n,\ n \in T\}$ by probabilistic laws. In Eq. 5.1, index n stands for time. This process is measured discrete in time and $T = \{0, 5, 10, \ldots\}$ years approximately. This time step is a reasonable time unit for land use change studies. Therefore, if the stochastic process considered a Markov process then the sequence of random variables will be produced by the Markov property, formally (Cabral and Zamyatin 2009):

$$P[X_{n+1} = a_{in+1} | X_0 = a_{i0}, \ldots, X_{in} = a_{in}] = P[X_{in+1} = a_{in+1} | X_{in} = a_{in}] \quad (5.1)$$

Table 5.1 Comparison of different accuracy assessment indices arising from diverse CA rules

Input file	Iteration number	Kappa index for built-up cells	Overall kappa	ROC value
1986 (predicted 1996)	2	0.6684	0.6953	0.837
	3	0.6684	0.6953	0.837
	4	0.7156	0.7000	0.846
	5	0.7546	0.6984	0.851
	6	0.7877	0.6925	0.854
	7	0.8151	0.6830	0.857
	8	0.8379	0.6705	0.86
	9	0.8562	0.6553	0.861
	10	0.8709	0.6382	0.859
1996 (predicted 2006)	3	0.6977	0.6912	0.825
	4	0.7475	0.6909	0.834
	5	0.7869	0.6838	0.84
	6	0.8201	0.6729	0.843
	7	0.8488	0.6598	0.847
	8	0.8736	0.6447	0.851
	9	0.8944	0.6280	0.854
	10	0.9116	0.6099	0.852

5.4.1 Markovian Property Test

Land use change in the study area needs to be proved as a Markovian process. In fact, it must have statistical dependence between X_{n+1} and X_n; and that statistical dependence is a first-order Markov process.

$$P(X_n = a_n | X_{n-1} = a_{n-1}) \neq P(X_n = a_n) \times P(X_{n-1} = a_{n-1}) \tag{5.2}$$

$$P[X_n = a_n | X_{n-1} = a_{n-1}] = P[X_n = a_n, X_{n-1} = a_{n-1}] / P[X_{n-1} = a_{n-1}] \tag{5.3}$$

A first-order Markov process is defined as a Markov process that the transition from one category to any other categories does not necessitate intermediate transitions to other states. The statistical dependence can be tested in any contingency table demonstrating the land cover changes between X_n and X_{n-1}. In this research, this test was performed for land cover changes between 1986–1996 and 1996–2006. To deduce from the association or independence between the land cover categories within the years from the contingency table, the random variable, with the chi-square distribution is defined by:

$$x^2 = \sum_i \sum_i \left((N_{ij} - M_{ij})^2 / M_{ij} \right) \tag{5.4}$$

Here, N is the contingency matrix showing the land cover change between two assumed time scales; for instance, either 1986–1996 or 1996–2006 or 1986–2006, and also, M the contingency matrix with the expected values of change, assuming the independence hypotheses.

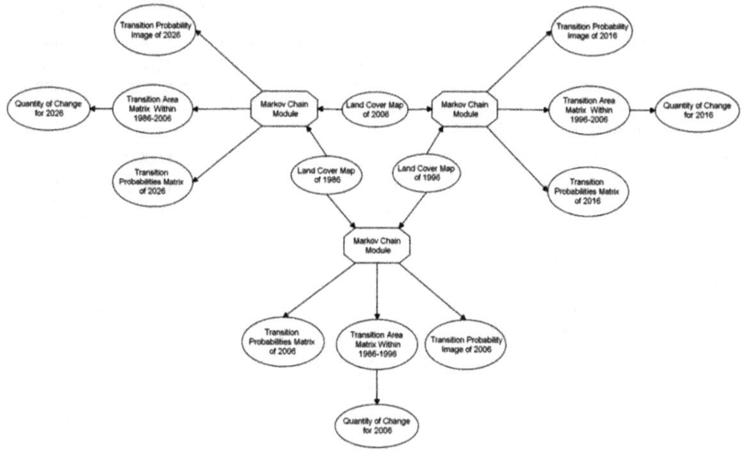

Fig. 5.7 Schematic view of the Markov chain model approach

x^2 basically measures the distance between the actual values of land cover change and the projected ones, assuming independence hypothesis and accordingly must be high enough to verify. The same non-parametric test was performed to assess the Markovian property. Thus, the values have to be compared with the observed values computed with the Chapman–Kolmogorov equation, supposing that these variables are generated by a first-order Markov process:

$$P(X_n = a_n|X_m = a_m) = P(X_1 = a_1|X_m = a_m) \times P(X_n = a_n|X_1 = a_1), \quad m \leq 1 \leq n \tag{5.5}$$

The Chapman–Kolmogorov equation expresses that the probability of transition between 1986 and 2006 can be projected by multiplying the transition probabilities matrix 1986–1996 by the transition probabilities matrix 1996–2006.

$$x^2 = \sum_i \sum_j \left((N_{ij} - o_{ij})^2 / o_{ij} \right) \tag{5.6}$$

5.4.2 Execution of the Markov Chain Module

The transition probabilities matrix is calculated by the contingency matrix displaying the relative frequencies of land change at a certain time period (Cabral and Zamyatin 2009). The IDRISI MARKOV module inputs a pair of land-cover images and outputs a transition probability matrix, a matrix of transition areas, as well as a set of conditional change probability images. A text file records the probability matrix that each land cover category will change to other categories under a certain probability value.

Table 5.2 Markov transition probabilities matrix between 1986–1996, 1996–2006 and 1986–2006

		Agricultural field	Built-up	Open land	Public park	Water body
Probability value of 2006 based on transition matrix of 1986–1996	Agricultural field	0.8835	0.0487	0.0615	0.0062	0.0001
	Built-up	0.0007	0.9907	0.0054	0.0031	0.0001
	Open land	0.0133	0.0689	0.9124	0.0052	0.0001
	Public park	0	0.0335	0.0232	0.9428	0.0005
	Water body	0.0105	0	0	0	0.9895
Probability value of 2016 based on transition matrix of 1996–2006	Agricultural field	0.9361	0.0402	0.0218	0.0017	0.0002
	Built-up	0.0036	0.9873	0.0055	0.0035	0
	Open land	0.0144	0.0576	0.9223	0.0055	0.0003
	Public park	0.0018	0.0088	0.0064	0.9816	0.0013
	Water body	0.0211	0.0102	0	0.0066	0.9621
Probability value of 2026 based on transition matrix of 1986–2006	Agricultural field	0.8469	0.0936	0.0493	0.0096	0.0005
	Built-up	0.0003	0.9885	0.0059	0.0052	0.0001
	Open land	0.0188	0.1131	0.8579	0.0099	0.0003
	Public park	0	0.0434	0.0198	0.9362	0.0005
	Water body	0.0105	0.0009	0	0	0.9886

The transition area matrix is a table which records the amount of pixels that are anticipated to change from one land cover category to other category according to a number of time units. The produced results (i.e. matrices) arising from this implementation were stored for use in further change analyses. This output determines the estimated quantity of change that can be used for the process of change allocation. Figure 5.7 presents a schematic view of the implementation of the Markov chain scenario.

In effect the Markov chain is not a spatially explicit model; therefore the Markov chain is not an appropriate model to estimate the location of change, which is the aim of GIS projects. Nevertheless, it is an excellent quantity estimator (Kamusoko et al. 2009) such that its outcomes can be allocated by means of other approaches. As is shown in Table 5.2, the probability of converting each land category to the others can be determined by the Markov chain model.

5.5 Cellular Automata Markov Scenario

This section of the chapter aims, in particular, to depict the cellular automata Markov model and how this module was executed. The cellular automata Markov model that has been designed into the IDRISI software (Andes Version) is an extension of multi criteria evaluation procedure which combines CA and Markov chain modules. By using the quantity of change which is calculated through the

Markov chain analysis (i.e. transition area matrix) the cellular automata Markov model applies a contiguity kernel to 'grow out' a land use map to a later time period; hence, this approach converts the outcomes of the Markov chain model to a spatially explicit model by integration of CA functionality. The certainty and accuracy of this module will be examined and demonstrated (see Fig. 5.8).

Some efforts were performed to construct high-resolution regional models by integration of the Markov and CA approaches (Clark 1990), and investigations in this area have been growing extensively (Wegener 2001). The Markov cellular automata model is a robust approach in terms of quantity estimation as well as spatial and temporal dynamic modelling of land use/cover changes, because GIS and remote sensing data can be capably incorporated. Biophysical and socioeconomic data could be used, firstly, to define preliminary conditions; secondly, to parameterise the Markov cellular automata model; thirdly, to analyse transition probabilities and, finally, to determine the neighbourhood rules with transition potential maps (Kamusoko et al. 2009). In the cellular automata Markov model, the Markov chain process manages temporal dynamics among the land use/cover categories based on transition probabilities, while the spatial dynamics are controlled by local rules determined either by the cellular automata spatial filter or transition potential maps (Maguire et al. 2005). In fact, the cellular automata Markov model begins allocating changes from the nearest cells to each land use type (Pontius and Malanson 2005).

In this section, the future land use/cover changes (up to 2026) in the study area were simulated based on the cellular automata Markov model, which combines Markov chain analysis and cellular automata models in order to change the essence of the Markov chain to a spatially explicit model.

The spatial resolution of output maps was defined at 30 m in accordance with Landsat imagery spatial resolution. The original cell size could avoid further uncertainty by employing reclassification functions. Hence, the quantity and percentage of each type of land use maps was calculated for the period of 1986–2006 in accordance with cross tabulation analysis.

5.5.1 Execution of the Cellular Automata Markov Model

Markov chain models have been broadly used to model land use changes including both urban and rural areas at coarse spatial scales. After preparing land use maps, transition probability matrices for both time periods were calculated as well as Markovian conditional probability images in IDRISI software (See Table 5.1).

The first record of Table 5.1 identifies the next 10-year step (i.e. 2006) as a description of transition probability matrix, where the agricultural areas category will remain at the same category at 88.35% probability and 4.87% will be converted to built-up area category. Furthermore, the value of the fourth row which identifies the probability of converting public parks category to agricultural land category is zero; in other words, it is not expected to observe any public park

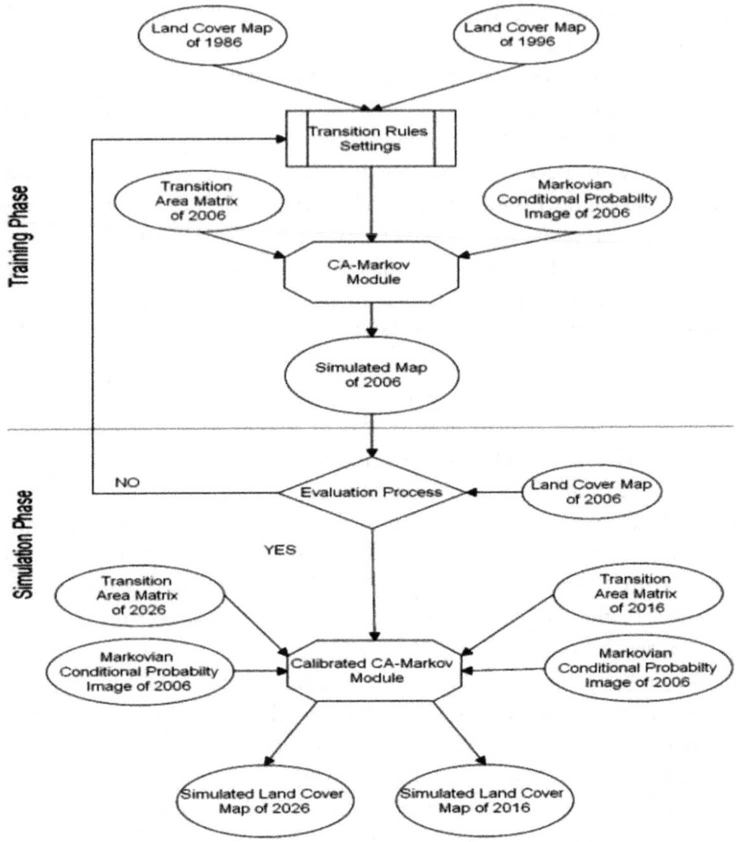

Fig. 5.8 Flowchart of the cellular automata Markov simulation process

cell that has been converted to agricultural field cells. Figure 5.9 demonstrates simulated maps arising from the implementation of the cellular automata Markov model at different iteration numbers.

The next step requires the need to set up the cellular automata Markov model for predicting the land use map. Since this module has Markovian property and CA behaviour, the cellular automata Markov model must be defined for both properties. Hence, by inputting the land use map of 1986, Markov transition areas parameters and transition suitability image parameters for Markovian property of the model were employed, as well as filter contiguity definition and number of iterations in support of cellular automata behaviour.

As shown in Fig. 5.8, it is aimed to simulate multiple land use maps for one-time step (e.g. 2006) by defining different transition rules. The simulated maps will be compared with the actual map, which allows us to evaluate the validity of this approach. Therefore, the verified model can be used to simulate future years.

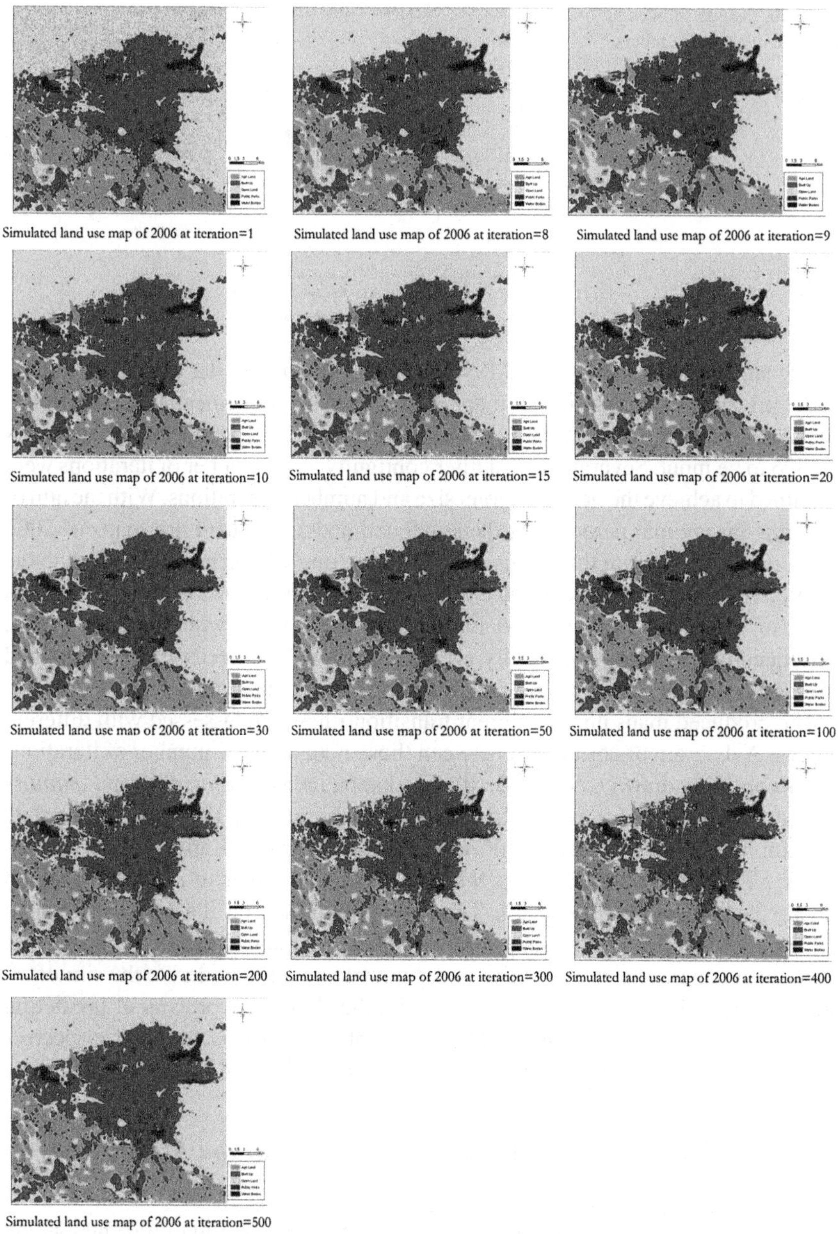

Fig. 5.9 Simulated land use map of 2006 from land use map of 1996 at different iteration numbers

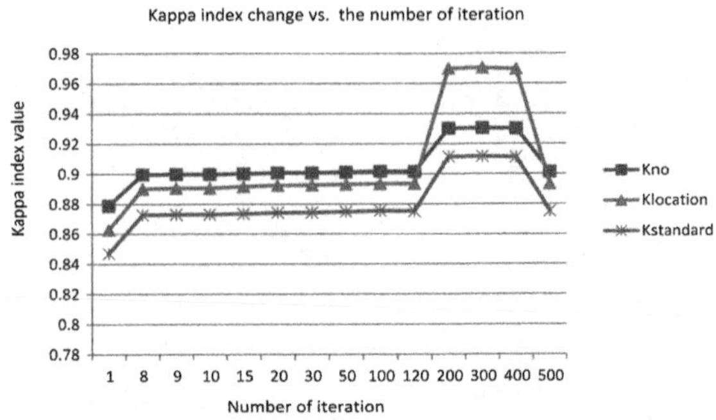

Fig. 5.10 Kappa indices at different iteration numbers

The land use maps of 1986 and 1996 were input to the cellular automata Markov model to produce a simulated map of 2006. This implementation requires a Markovian conditional probability image of 2006 and, also, a transition area matrix of 2006 to be input. Several types of filter contiguity and a number of iterations were examined to achieve the optimal kernel size and number of iterations. With the aim of reaching the optimal parameters, the simulated and actual land use maps of 2006 were crossed to validate the results. One of the setting parameters was to define the iteration number that will reproduce different maps. This model evaluation process needs to verify all the simulated maps to compare them with the actual map; consequently, the most statistically similar map will be selected. The predefined parameters will be chosen as the proper settings for the next runs.

The produced maps under different transition rules were assessed with different indices. A diagram of correlation between those maps and the number of iterations was accordingly drawn (see Fig. 5.10). The kappa indices of *location* and *quantity* for the simulated maps were calculated, and subsequently the most appropriate iteration number at iteration of 300 was determined with a Kappa standard index of 0.91. The input transition rules were considered in order to run this approach and predict future land use maps. This was done based on the transition probabilities matrices of land change (1996–2006) and land change (1986–2006). Markovian conditional probability images have to be input to derive the simulated land use maps of 2016 and 2026. Eventually, the simulation process of predicting the land use maps of 2016 and 2026 was implemented to output the respective maps. These maps are represented in Chap. 7 (Figs. 7.3, 7.4).

5.5.2 Validation of the Cellular Automata Markov Model

A cross comparison between the simulated maps at different iterations and actual maps was employed to verify the certainty of the model. The highest value of accuracy among the resultant maps was chosen, which is approximately 91% for the kappa index, and 97% for K-Location (Fig. 5.11). Investigation of this model shows that the cellular automata Markov model is a good estimator for the quantification of change and continuous-space change modelling. Based on visual analysis, this model

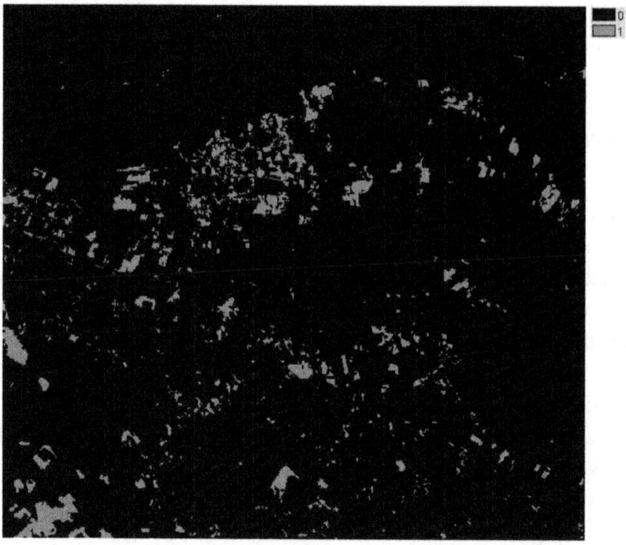

Fig. 5.11 Dependent variable *Y*; change to built-up area between 1986 and 1996 (no change: $Y = 0$; change $Y = 1$)

produces some diffused-speckle developed cells which do not correspond with the reality. Besides, this model needs a lot of time to run the simulation process and, also, to be replicated for a huge number of iterations (e.g. 300, 400). Although the simulated maps have high Kappa indices the edges of land categories appear wavy and circular in shape, which do not match with the reality, i.e. they seem unreal.

5.6 The Logistic Regression Model Scenario

The logistic regression analysis has been the most frequently used approach during the past two decades for predictive modelling by means of variation of inductive modelling (Verhagen 2007). Empirical estimation and dynamic simulation models have been used to simulate land use/cover changes. Various types of rule-based modelling (e.g. cellular automata model) are the most suitable models for incorporating spatial interaction effects and handling temporal dynamics. CA models, however, focus primarily on the simulation of spatial patterns rather than the interpretation of spatio-temporal processes of urban sprawl. There is a lack of incorporation between most dynamic simulation models over socioeconomic variables (Hu and Lo 2007). In this section, another approach by means of the logistic regression model on urban sprawl will be explained. The aim of executing this technique was to observe the presumed relationship and interactions between social, economic and environmental parameters which could drive urban expansion. As far as it has been realised, this technique has never been published or even employed upon the study area. Hence, this implementation and its outcomes could lead to more accurate results in this area of research, and achieve a better understanding of the interaction between those variables.

5.6.1 An Overview of the Logistic Regression Technique

Regression is a method to discover the coefficients of the empirical relationships from observations. Linear regression, log-linear regression and logistic regression are the most used regression approaches (Hu and Lo 2007). In logistic regression, the dependent variable can be either binary or categorical, and the independent variables could be a set of categorical and continuous variables. Routine assumption is not required for the logistic regression model. Hence, logistic regression is advantageous in comparison with the linear regression or log-linear regression. It is fundamental to extract the coefficients of independent variables from the observation of land use conversion, since urbanisation does not frequently follow typical supposition, and its prominent factors are usually a combination of continuous and categorical variables (Xie et al. 2005). The general form of logistic regression is as follows:

$$y = a + b_1 x_1 + b_2 x_2 + \cdots + b_m x_m \tag{5.7}$$

$$y = \log_e\left(\frac{P}{1-P}\right) = \log\ it\,(p) \tag{5.8}$$

$$P = \frac{e^y}{1+e^y} \tag{5.9}$$

Where $x_1, x_2, \ldots, x_m$ are independent variables, *y* defines a linear combination function of the independent variables representing a linear relationship. Moreover, the $b_1, b_2, \ldots, b_m$ parameters are the regression coefficients to be retrieved. Function y is known aslog it (*P*) i.e. the logarithm (base-e) of the odds or likelihood ratio that the dependent variable *Z* is 1. Probability value (*P*) strictly increases while *y* value goes up. Regression coefficients b_1 to b_m imply the contribution of each independent variable on the probability value. A positive value implies that the independent variable helps to increase the probability of land change and a negative value implies the reverse effect. The statistical method is a multivariate estimation process which examines the relative significance and strength of the factors. While employing logistic regression to simulate rural–urban land transformation, it is crucial to consider the spatial heterogeneity of spatial data. Spatial statistics such as spatial dependence and spatial sampling also have to be taken into account to eliminate spatial autocorrelation (Hu and Lo 2007). Otherwise, unreliable factor estimation or unproductive estimates (i.e. wrong results) of the hypothesis test will be produced.

There are two basic approaches to assess spatial dependence: firstly, building a more complex model incorporating an autoregressive structure and, secondly, designing a spatial sampling plot to enlarge the distance interval between sampled points. Spatial sampling creates a smaller sample size that loses certain information and conflicts with the large sample of asymptotic normality of maximum likelihood method, upon which logistic regression is based on. Nonetheless, it is a reasonable approach to eliminate spatial auto-correlation, and a reasonable design of spatial sampling scheme will make an ideal balance between the two sides (Xie et al. 2005).

The logistic regression model is employed to predict a categorical variable from a set of predictor variables. A discriminated function analysis is generally employed if all of the predictors are continuous and properly distributed; Logit analysis is generally utilised if every predictor is categorical. In fact, logistic regression is often preferred if the predictor variables are a set of categorical and continuous variables. Besides, they should be properly distributed. The predicted dependent variable in a Logistic Regression Model is a function of the probability that a particular theme will be in one of the categories; for instance, the probability of change upon a specific land use based on a set of scores on the predictor variables such as proximity to interchange network, and so on (Huang et al. 2009).

LOGISTICREG module in IDRISI Andes performs binomial logistic regression, in which the input dependent variable must be binary in nature and can have only two possible values (0, 1). Such regression analysis is usually employed in the estimation of a model that depicts the relationship between continuous independent variables to a binary dependent variable. The basic assumption is that the probability of a dependent variable takes the value of 1 (positive response). The logistic curve and its value can be calculated with the following formula: (Mahiny and Turner 2003)

$$P(y=1|X) = \frac{\exp(\sum BX)}{1+\exp(\sum BX)} \tag{5.10}$$

Where:

P is the probability of the dependent variable occurrence

X is the independent variables, $X = (x_0, x_1, x_2 \ldots x_k), \quad x_0 = 1;$

B is the estimated parameters, $B = (b_0, b_1, b_2 \ldots b_k)$

In order to linearize the above model, as well as remove the 0/1 boundaries for the original dependent variable which is probability, the following transformation is usually applied:

$$\mathrm{P}' = \mathrm{In}(p/(1-p)) \tag{5.11}$$

This transformation is referred to as the Logit or logistic transformation. Thus, after the transformation P′ can theoretically assume any value between plus and minus infinity (Hill and Lewicki 2007). By performing the Logit transformation on both sides of the above Logit regression model, we obtain the standard linear regression model:

$$\mathrm{In}(p/(1-p)) = b_0 + b_1 \times x_1 + b_2 \times x_2 + \ldots + b_k \times x_k + \text{error\_term} \tag{5.12}$$

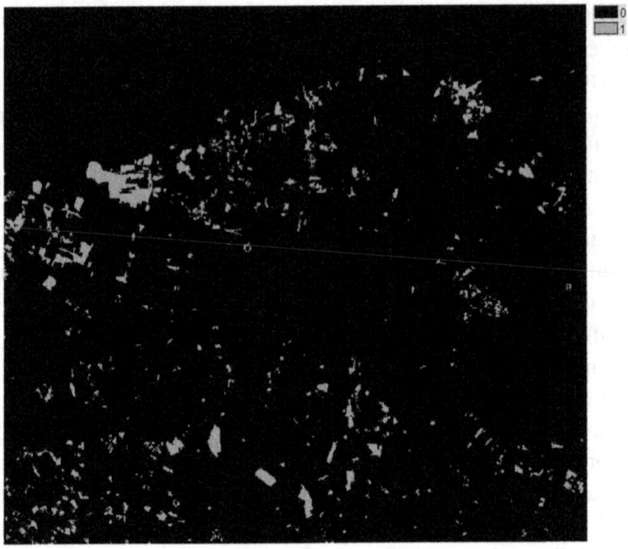

Fig. 5.12 Dependent variable Y; change to built-up area between 1996 and 2006 (no change: $Y = 0$; change $Y = 0$

In fact the Logit transformation of binary data ensures that the dependent variable will be continuous, and the new dependent variable (Logit transformation of the probability) is boundless. Furthermore, it ensures that the probability surface will be continuous within the range from 0 to 1. In general, systematic sampling and random sampling are two approved sampling methods in logistic regression. Systematic sampling reduces spatial dependence. On the other hand, random sampling is capable of representing population, but does not efficiently reduce spatial dependence, especially local spatial dependence (Huang et al. 2009).

5.6.2 *Implementation of the Spatially Explicit Logistic Regression Model*

In this section of this chapter, it is intended to clarify the assumed independent and dependent variables and the interactions between these variables. Also, a description over model validation and outputs will be presented and simulated maps of future years will be demonstrated. Accordingly, we start with the identification of dependent and independent variables, and then the effective factors upon the dependent variable will be depicted.

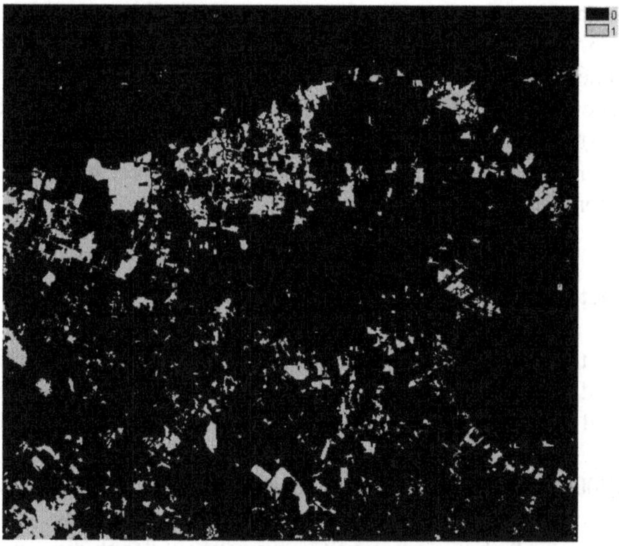

Fig. 5.13 Dependent variable *Y*; change to built-up area between 1986 and 2006 (no change: $Y = 0$; change $Y = 1$)

Table 5.3 ROC and adjusted odd ration values for 18 sets of variables

	ROC	Adjusted odd ratio
Variables set 1	0.8441	20.102
Variables set 2	0.7831	7.6964
Variables set 3	0.844	21.7224
Variables set 4	0.7766	5.2355
Variables set 5	0.6635	3.0513
Variables set 6	0.9223	26.2327
Variables set 7	0.9352	50.3255
Variables set 8	0.9218	26.0128
Variables set 9	0.7167	3.2804
Variables set 10	0.7187	3.4114
Variables set 11	0.8906	16.052
Variables set 12	0.8915	16.5333
Variables set 13	0.8945	15.942
Variables set 14	0.7531	4.7522
Variables set 15	0.8031	11.586
Variables set 16	0.8053	12.0991
Variables set 17	0.8039	11.4385
Variables set 18	0.7392	5.7809

5.6.2.1 Identification of the Dependent Variable

The dependent variable in this implementation is the quantity of change from no-built-up area to built-up area presented as a binary raster lattice where value 1 introduces change on the specific pixels and zero indicates no-change pixels. Figures 5.11, 5.12, and 5.13 represent the structure of the dependent variable files.

A set of independent variables was imported to the Logistic Regression Model in order to become self-calibrated, with the support of IDRISI Andes GIS software (see Table 5.3). A defined mask upon all input data was employed at 30 m resolution to create equal dimension raster files; however, it was an intensive computation for the computer hardware.

5.6.2.2 Predictor Variables (Independent Variables)

In this section, the prior produced land use maps for the years 1986, 1996 and 2006 were employed to specify the change over built-up areas between 1986–1996, 1996–2006 and 1986–2006. Logistic regression modelling executes a data-driven rather than a knowledge-based approach in picking the predictor variables (Hu and Lo 2007). A set of predictor variables was chosen based on preliminary investigations over the case study as well as expert knowledge. A review of effective variables, which was employed in previous similar studies, was a helpful guide. Statistical evaluation, retrieving ROC values and adjusted odd ratios for each set of variables were investigated to pick the optimum set (see Table 5.2). Thus, a calibration process needed to be utilised in order to assure the effectiveness of the assumed variables. These variables and process of data compilation will be explained in the next section.

5.6.2.3 Data Compilation

The social variables correspond to the four affordable elements shaping Tehran's urban patterns (population density, distance to building blocks, single building features, farming lands, categorical demography). Other social variables data were not accessible to be utilised in this approach. Population density is a social variable which determines per capita population per area unit and is expressed as persons per hectare. The econometric and biophysical variables correspond to the eleven affordable elements shaping Tehran's metropolitan patterns (distance to CBD; distance to nearby cities; distance to road networks and interchange; open land features; easting and northing coordinate; digital elevation model; park features; distance to stream; and slope) (Hu and Lo 2007). A set of independent variables (X_1–X_{17}) was imported to a logistic regression model, supported by IDRISI Andes software. An input dataset was designed at 30 m resolution due to compatibility with other available data. Although, a set of other input data, such as distance to education and administration areas, and distance to factories had been evaluated as input to the model, because of weak results, this input data were rejected; hence, these seventeen datasets were imported into the model (see Table 5.4).

Table 5.4 Dependent and independent variables in the logistic regression approach

	Variable	Denotation	Structure of variable
Dependent	Y	0—No change to built-up 1—Change to built-up	Dichotomous
Independent	X_1	1—Single building features 0—Non single building features	Binary
	X_2	Proximity to nearby cities (m)	Continuous
	X_3	Proximity to interchange (m)	Continuous
	X_4	1—Farming land features 0—Non farming land features	Binary
	X_5	1—Open land features 0—Non open land features	Binary
	X_6	Proximity to building blocks (m)	Continuous
	X_7	Easting coordinates (m)	Continuous
	X_8	Northing coordinates (m)	Continuous
	X_9	Proximity to CBD (m)	Continuous
	X_{10}	Proximity to road network (m)	Continuous
	X_{11}	Digital Elevation Model (m)	Continuous
	X_{12}	Population density (person/ha)	Continuous
	X_{13}	Park features	Binary
	X_{14}	Proximity to stream (m)	Continuous
	X_{15}	Slope (%)	Continuous
	X_{16}	Categorical demography	Categorical
	X_{17}	Proximity to residential districts	Continuous

Spatial correlation may exist between each category of variables so that logistic regression is able to drop the correlated variables according to the statistical calibration. This calibration basically checks for multi co-linearity. Model calibration in this study was done in two steps, including initial calibration and refining, respectively. All required data were converted to raster format at 30 m resolution.

5.6.3 Calibration of the Logistic Regression Model

The optimum set of variables was picked based on Table 5.2. Each set of variables had different ROC and adjusted odd ratio, which verified the validity of the model, and the approach was carried out numerous times. In order to select the optimum set of variables, it had to reach the highest ROC value. In fact, ROC = 1 indicates a perfect fit and ROC = 0.5 indicates a random fit. A higher adjusted odds ratio is expected for a better fit and higher validity. Therefore, the optimum set of variables is demonstrated in Fig. 5.14.

The logistic regression module was implemented 18 times for 18 sets of variables in order to reach the highest possible ROC and adjusted odd ratio values. The highest value of 0.9532 was obtained, which verifies the accuracy of this model.

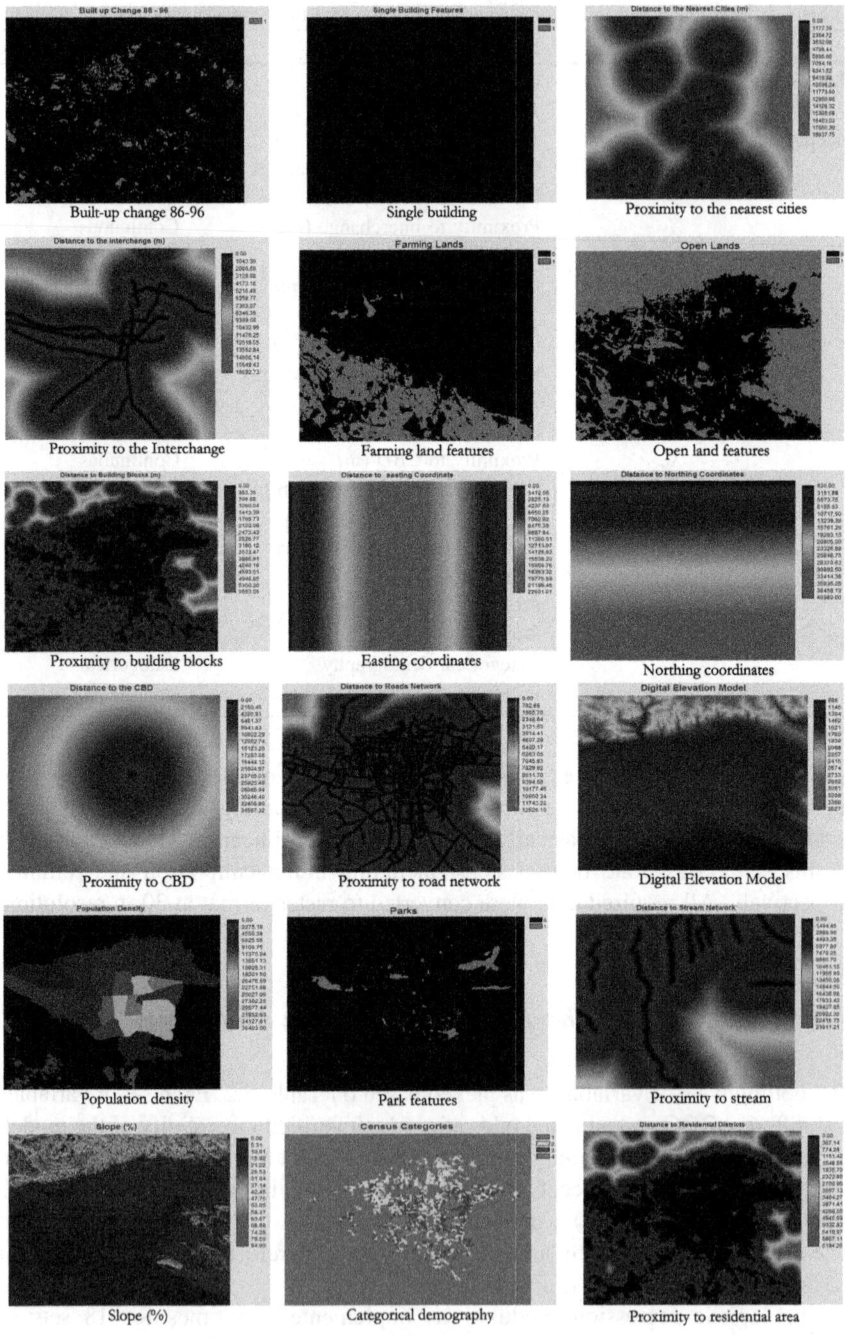

Fig. 5.14 Raster layers of independent variables represented in binary and continuous values

Furthermore, the optimum set of variables was incorporated in the model refining phase in order to correct any spatial autocorrelation that might exist. Thus, the selected combination had the minimum spatial autocorrelation. In Table 5.4, a descriptive table of appropriate variables, as well as their structure, is shown. The dependent variable (i.e. built-up change) and independent variables (X_1–X_{17}) are separated by assigned units in the mentioned table.

The employed data and the input maps are shown in Fig. 5.14. These maps are the ultimate variables which have been discussed previously.

The model produces an equation that shows the rate of effectiveness of each particular variable. This equation is presented in the following Eq. 5.13.

$$\begin{aligned}\text{Logit (Urban growth 86 − 96)} = & -23.1033 \text{ (intercept)} \\ & + 0.000165 \times \text{Proximity to CBD} \\ & + 0.597356 \times \text{Categorical demography} \\ & - 0.00001 \times \text{Proximity to nearby cities} \\ & - 0.000072 \times \text{Northing coordinates} \\ & + 0.000236 \times \text{Population density} \\ & - 7.428991 \times \text{Proximity to residential area} \\ & + 1.367012 \times \text{Proximity to single buildings} \\ & - 0.000061 \times \text{Easting coordinates} \\ & + 19.776172 \times \text{Farming lands} \\ & - 0.003773 \times \text{Proximity to building blocks} \\ & - 0.001391 \times \text{DEM} \\ & - 0.000044 \times \text{Proximity to interchange} \\ & + 20.618511 \times \text{Open lands} \\ & + 18.393214 \times \text{Proximity to parks} \\ & + 0.000026 \times \text{Proximityto roads} \\ & - 0.047149 \times \text{Slope} \\ & - 0.000013 \times \text{Proximity to streams}\end{aligned} \tag{5.13}$$

According to Eq. 5.13, some variables which have positive values are more favourable for development (e.g. proximity to the CBD, categorical demography, population density, proximity to single buildings, farming lands, open lands, proximity to parks, and proximity to roads). Where variables return negative values the attraction for development falls significantly (e.g. proximity to nearby cities, proximity to streams, northing coordinates, easting coordinates, proximity to residential area, proximity to building blocks, elevation, slope, and proximity to interchange). In other words, those pixels which are closer to the CBD area have more probability of development, and those cells which are in steep slopes have less probability of change. Importantly, the coefficients explain the intensity of

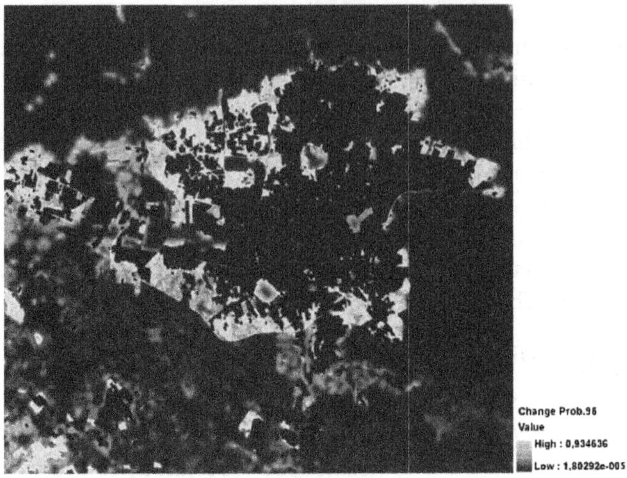

Fig. 5.15 Transition surface maps of study area for 1996

influence in the occurrence of development, for example, proximity to parks is a significant factor in such development.

The output product of the logistic regression model is a probability surface of dependent variable occurrence, which is in this approach urban development (see Figs. 5.15 and 5.16). The probability surface shows that each single cell will be developed with a particular amount of probability. However, this approach is not able to specify the amount and location of change, but can be integrated with other techniques to quantify and allocate the quantity of change. Hence, this probability map will be integrated with the Markov chain model to quantify the extent of the changes. Thereafter, the obtained quantity of change will be allocated in the entire map. The allocation process starts from the maximum value of probability working downward. This process will be explained in Chap. 7.

5.6.4 Validation of the Logistic Regression Model

By means of the prepared probability surface, the quantity of change can be specified through possible techniques, either the Markov chain model or by population growth estimation. The Markov chain model has already been explained in detail. The second method is to employ a footprint of inhabitants to reach the quantity of change (see Sect. 4.10). In this approach, the amount of change was determined based on the transition matrix of the Markov chain model to quantify the changes. The obtained amount was input to the allocation phase. A code was written in Python to subtract the existing built-up areas before beginning the allocation of change from the highest probable cell to the lowest probable cell.

Hence, after executing the designed logistic regression approach, a predicted transition probability surface map, and a residual map indicating the difference

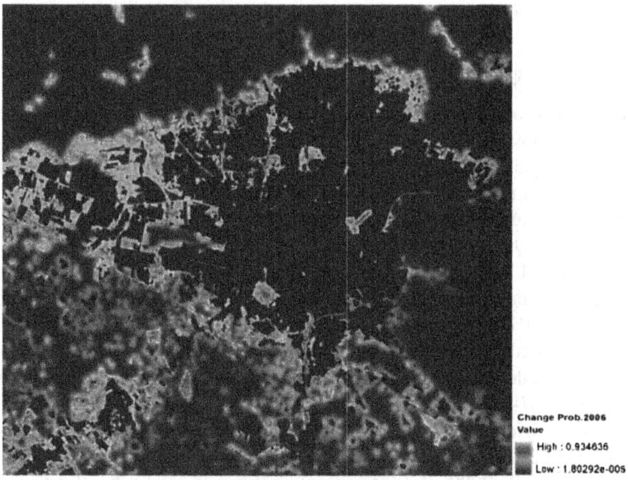

Fig. 5.16 Transition surface maps of study area for 2006

between the predicted and the observed probability, were achieved. Therefore, a transition surface map was produced for 2006 onward. The mentioned prediction surface maps are shown in Figs. 5.15 and 5.16, which can be used for change specification for upcoming periods (2016, 2026). This task was carried out and is demonstrated in Chap. 7.

5.6.5 Land Change Prediction

After the process of model validation was undertaken and the qualification of this model was ensured, land use maps were predicted for 2016 and 2026. Logistic regression requires updated data for the specific times to establish more accurate prediction. In other words, the actual road network map for 2016 is required for the creation of the probability surface at this juncture in time. Therefore, a multi temporal data set of the study area was gathered.

5.7 Summary

Several traditional techniques were demonstrated within this chapter (e.g. CA, Markov chain model, CA-Markov model, logistic regression model). Each model was firstly evaluated and validated and then, once assured of its performance, a land change map was predicted for two future time steps (i.e. 2016, 2026). Each model had some advantages and disadvantages which were investigated, and will be discussed in Chap. 7. The intention was to gather these results in order to integrate them into the assumed ABM. In the next chapter, we start designing the ABM model based on results emanating from traditional methodologies as well as ABM characteristics.

References

Batty M, Xie Y (1994) From cells to cities. Env Plann B: Plann Des 21(7):31–48

Cabral P, Zamyatin A (2009) Markov processes in modeling land use and land cover changes in Sintra-Cascais, Portugal. Dyna 76(158):191–198

Clark JD (1990) Modeling and simulating complex spatial dynamic systems: a framework for application in environmental analysis. SIGSIM Simul Dig 21(2):9–19

Couclelis H (1985) Cellular worlds: a framework for modeling micro–macro dynamics. Env Plann A 17(5):585–596

Goodchild MF, Steyaert LT, Parks BO, Johnston C, Maidment D, Crane M, Glendinning S (1996) GIS and environmental modeling: progress and research issues. Wiley, New York

Hill T, Lewicki P (2007) Statistics methods and applications. StatSoft. Tulsa, OK. http://www.statsoft.com/textbook/neural-networks/#linear

Hu Z, Lo C (2007) Modeling urban growth in Atlanta using logistic regression. Comput Env Urban Syst 31(6):667–688

Huang CY, Sun CT, Hsieh JL, Lin H (2004) Simulating SARS: small-world epidemiological modeling and public health policy assessments. J Artif Societies Soc Simul 7(4)

Huang B, Zhang L, Wu B (2009) Spatiotemporal analysis of rural-urban land conversion. Int J Geog Inf Sci 23(3):379–398

Kamusoko C, Aniya M, Adi B, Manjoro M (2009) Rural sustainability under threat in Zimbabwe—simulation of future land use/cover changes in the Bindura district based on the Markov—cellular automata model. Appl Geogr 29(3):435–447

Li X (2008) Simulating urban dynamics using cellular automata. In: Liu L, Eck J (eds) Artificial crime analysis systems: using computer simulations and geographic information systems, pp 125–139

Li W, Li B, Shi Y (1999) Markov-chain simulation of soil textural profiles. Geoderma 92(1–2): 37–53

Liu Y (2008) Modelling urban development with geographical information systems and cellular automata, 1st edn. CRC Press, Boca Raton, FL

Maguire D, Batty M, Goodchild M (2005) GIS, spatial analysis and Modeling. Esri Press, Red lands, CA

Moreno N, Wang F, Marceau DJ (2009) Implementation of a dynamic neighborhood in a land-use vector-based cellular automata model. Comput Env Urban Syst 33(1):44–54

Pontius RG Jr, Malanson J (2005) Comparison of the structure and accuracy of two land change models. Int J Geog Inf Sci 19(2):243–265

Salman Mahiny AS, Turner BJ (2003) Modeling past vegetation change through remote sensing and GIS: a comparison of neural networks and logistic regression methods. In: Proceedings of the 7th international conference on geocomputation, University of Southampton, United Kingdom

Verhagen P (2007) Case studies in archaeological predictive modeling. Leiden University Press, Leiden

Wegener M (2001) New spatial planning models. Int J Appl Earth Obs Geoinf 3(3):224–237

White R, Engelen G (1993) Cellular automata and fractal urban form: a cellular modelling approach to the evolution of urban land-use patterns. Env Plann A 25(8):1175–1199

Xie C, Huang B, Claramunt C, Chandramouli C (2005) Spatial logistic regression and gis to model rural-urban land conversion. In processus second international colloquium on the behavioural foundations of integrated land-use and transportation models: frameworks, models and applications. University of Toronto, Canada

Yang Q, Li X, Shi X (2008) Cellular automata for simulating land use changes based on support vector machines. Comput Geosci 34(6):592–602

Chapter 6
Designing and Implementing Multi Agent Geosimulation

6.1 Introduction

The aim of this section is to bring an overview of the designed steps of ABM implementation, followed by more in-depth detail. This scenario begins with the classification of the agents according to their significant influences. The strengths of traditional methods will be imported into the presumed ABM model to increase the accuracy of the results, the outcomes of which will be depicted in the next chapter in order to compare the separate models. It was our intention to design an ABM within the GIS environments by means of GIS functionalities and coding environments, but in fact, the ArcGIS software and Python were used to implement this model.

6.2 Abstract Model of the ABM

In this section, an abstract model of the ABM scenario will be depicted in detail. Based on domestic land policies in Tehran, it was assumed that three agents are in control of the issues of land change in the study area. These agents are resident agents, developer agents, and government agents. According to similar studies carried out in Chinese cities, these agents interact with each other to develop land for built-up purposes. For instance, the resident agent has some preferences for choosing a particular place to live, which can hinge on a couple of factors (e.g. proximity to some infrastructures and resources as well as natural environment). Elsewhere, there are some developers in the city perimeter who are more interested in financial profit from land development. Hence, these developers look for affordable places where they can earn more money after their initial investment. In order to have a successful investment, they supply some of the requirements for settling new inhabitants, in areas which are more popular for

J. Jokar Arsanjani, *Dynamic Land-Use/Cover Change Simulation: Geosimulation and Multi Agent-Based Modelling*, Springer Theses,
DOI: 10.1007/978-3-642-23705-8_6, © Springer-Verlag Berlin Heidelberg 2012

first-time affordable home seekers to rent or buy property. Hence, these two agents satisfy, to a degree, a mutual interest. Later in this chapter, this interaction will be covered in more detail.

Also, the aim of this exercise was to establish the optimum weights associated to each particular variable. Whilst the developer agents decide to utilise land for property and profit motives, official approval is required for the project in a given area. If the target location has no constraints for development, then the application will be approved automatically, otherwise the government agent will refuse the application. Consequently, in the development process, these three agents are together involved in whether to allow this change to happen or not. These agents and their distribution will be discussed further in this chapter.

6.3 Agents Characteristics and Behaviour

Three major types of agents were considered to take all effective variables into account in this approach. In this model, each agent represents behaviours expected of that particular agent. Each type of agent has unique features. For instance, the government agent has the ability to define the proper sites for development and avoid or reject unsuitable sites for development and, also, the power to protect specific areas (e.g. national parks). The location of the developer agent was another concern for modelling, but the primary assumption was to take into account that each cell at the raster space of the study area is maintaining a set of agents; i.e. each cell contains a resident, developer and government agent.

Therefore, each agent can act independently and simultaneously and their interactions can cause the final decision of the mentioned agents. Resident agents are mobile; however, in this study area, the developer agent is leading the resident agents where to settle. In other words, whereas the developer agent affords and constructs new housings and neighbourhoods, it is the resident agent who selects their favourite locations, based on affordability and other criteria (e.g. accessibility to required facilities). This issue will be discussed in the following sections.

It is vital to define the behaviours of the predefined agents appropriately to gain better results. It would be more robust to label agents' decision behaviours geospatially explicit, where each major type of agent could be categorised into subtypes according to the agents' properties and characteristics. For instance, resident agents can be classified into multiple groups based on incomes, household size and other proper variables. Each collection of residents has unique preferences of choosing proper sites for settlement. By classification and simplification of this task, a vast decrease in computation time will be reached. This can help to avoid any autocorrelation. The defined agents will influence each other in making decisions. Resident agents decide to live in the suitable sites, where they can gain their preferences such as accessibility to the road networks, shopping malls, parks, leisure facilities, public transport, etc.

Therefore, property developers have to alter their investment policies and plans according to the residents' purchasing behaviour. Their simple objective is to earn profit as much as possible. However, the developers must get the appropriate approval from the government before any place becomes developed. Government is the final decision maker who is able to approve the accuracy of site development by considering the environmental circumstances and internal policies. In fact, if a lot of buyers exist for selecting the same locations, housing prices will rise up and developer agents will construct new and intensive housings, therefore the vicinity of that neighbourhood will be developed. Thus, if the purchase or rent prices exceeds their affordable threshold, they have to look for other places for residency.

6.4 Spatial Distribution of the Agents

As mentioned previously, five land categories exist in the land use maps: agricultural lands, built-up areas, open lands, public parks and water bodies. Each single cell has the potential to be developed, i.e. any pixel of those land classes can be converted to built-up areas provided the local government issues permission for development. Hence, we did not exclude any area in the manipulation process. All agents are distributed equally in the entire extent inside the 30 m pixels. This means that in each pixel of the study area, those three agents coexist. Thus, these agents must collaborate in the model when any change takes place. This means that if the resident agent and developer agent inside a pixel meet their common interests, then the government agent acts in a binary form.

6.5 Classification of Agents

In this section of the chapter, it is intended to classify the effective agents in land change matter in the study area. This classification is made based on characteristics of the change drivers; for instance autonomy, their major effects, and the importance of their decisions. Therefore, the following agents were created in order to define their behaviours and interactions. Three main agents were assumed for the multi-agents simulation: government agents, developer agents, and resident agents.

6.5.1 Resident Agents

In the study area, two kinds of residents exist: those who are moving into the metropolitan area from other cities to live and, secondly, current residents who relocate to the better or inferior places due to their financial situation.

The behaviours of both residents can influence the type of change and lead the developer agents. Furthermore, these behaviours can justify the investment plans for developer agents. These resident agents and their interactions are the main keys for the formation of urban expansion.

Several variables already were picked for logistic regression implementation. The effectiveness of these variables was satisfied by the model statistics in Sect. 5.6. These variables play a key role for resident agents to choose an area for settlement, which are as listed below;

- Favourite elevation
- Favourite slope
- Accessibility to medical services
- Accessibility to metro stations
- Distance from disposal areas
- Accessibility to orchard areas
- Accessibility to sport centres
- Accessibility to road networks (paved roads, freeways, highways, roads)
- Accessibility to recreation points
- Accessibility to commercial centres
- Distance to railways

In order to consider all possible behaviours of resident agents, multiple factors were taken into account to maximise the efficiency of the model. A utility function was designed for resident agents to control all possible variables based on their effectiveness, i.e. applying each variable based on its impact which will be discussed in the following notes (see Fig. 6.1). The main purpose is to maximise the accuracy of the change allocation model. Therefore, a utility function of location (ij) for resident agent k can be demonstrated as in Eq. 6.1.

$$\begin{aligned} F(k, ij) = {} & W_{\text{education}} B_{\text{education}} + W_{\text{elevation}} B_{\text{elevation}} + W_{\text{slope}} B_{\text{slope}} \\ & + W_{\text{medical}} B_{\text{medical}} + W_{\text{metro}} B_{\text{metro}} + W_{\text{disposal}} B_{\text{disposal}} \\ & + W_{\text{orchard}} B_{\text{orchard}} + W_{\text{sport}} B_{\text{sport}} + W_{\text{road}} B_{\text{road}} \\ & + W_{\text{recreation}} B_{\text{recreation}} + W_{\text{commercial}} B_{\text{commercial}} \\ & + W_{\text{railway}} B_{\text{railway}} + \varepsilon_{tij} \end{aligned} \tag{6.1}$$

Where $W_{\text{education}} + W_{\text{elevation}} + W_{\text{slope}} + W_{\text{medical}} + W_{\text{metro}} + W_{\text{disposal}} + W_{\text{orchard}} + W_{\text{sport}} + W_{\text{road}} + W_{\text{recreation}} + W_{\text{commercial}} + W_{\text{railway}} = 1$ and define the weight of each particular factor and the summation of them is 1. The other parameters denoted by B (e.g.B_{railway}) are the abstracted form of some factors which are addressed in Table 6.1.

The mentioned factors in Table 6.1 were input into the designed function in order to retrieve their weights. These weights will be retrieved through an AHP method execution. These variables will be calculated for resident k and location (ij). Here, ε_{tij} is a stochastic term. A descriptive table of achieved weights is shown in Table 6.2 (see Sect. 6.5.1.2).

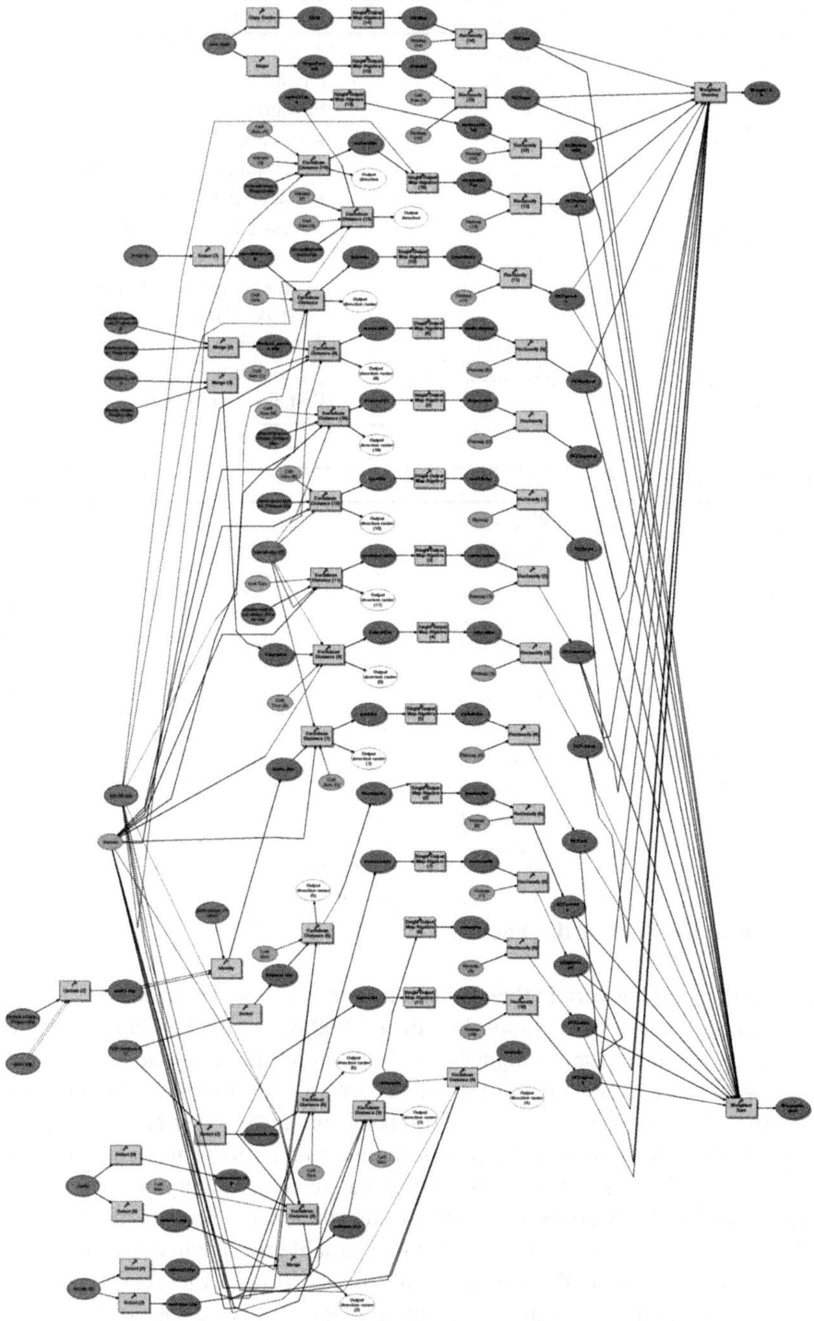

Fig. 6.1 Schematic view of the designed geospatial model for the resident agents

Table 6.1 Effective factors in controlling resident agent preferences for settlement

Factors	Denotation
$B_{\mathrm{education}}$	Accessibility to education centres
$B_{\mathrm{elevation}}$	Favourite elevation
B_{slope}	Favourite slope
B_{medical}	Accessibility to medical services
B_{metro}	Accessibility to metro stations
B_{disposal}	Distance from disposal areas
B_{orchard}	Accessibility to orchard areas
B_{sport}	Accessibility to sport centres
B_{road}	Accessibility to road networks (paved, freeways, highways, roads)
$B_{\mathrm{recreation}}$	Accessibility to recreation points
$B_{\mathrm{commercial}}$	Accessibility to commercial centres
B_{railway}	Distance to railways

Table 6.2 Weights of the effective factors upon the resident agent

Variable	Weight
$W_{\mathrm{education}}$	0.0242
$W_{\mathrm{elevation}}$	0.0125
W_{slope}	0.1758
W_{medical}	0.0435
W_{metro}	0.0664
W_{disposal}	0.0237
W_{orchard}	0.0573
W_{sport}	0.0794
W_{roads}	0.3107
$W_{\mathrm{recreation}}$	0.0871
$W_{\mathrm{commercial}}$	0.019
W_{railway}	0.1004

6.5.1.1 Fuzzification of the Factors

Each noted factor has certain thresholds for their favourite and undesired domains. For instance, areas with elevation less than 1,600 m are the favourable locations for residency, whereas areas higher than 2,000 m are not affordable for housing. These areas have some difficulties for settlement due to some meteorological situations. The areas with elevation in between have a degree of vagueness. A fuzzification process could help us to define fuzzy sets. Thus, applying fuzzy membership functions seems to be an innovative tool, which can help to reach optimum outcomes. Certain thresholds were applied for each particular factor. A fuzzy membership function for each single variable was applied to classify those variables' effectiveness. The fuzzification process allows us to create categorical variables. This would be helpful in defining rules to allocate specified changes.

The utility function basically affects the location behaviours of resident agents. In Eq. 6.1, if those variables were manipulated evenly, all the weights could be assigned with equal values. In other words, each variable in this equation takes an approximate value at 0.0833. Because the resident agents have different preferences in choosing favourite locations for settlement, this can be considered by implementing appropriate weights in the utility function. Therefore, a higher value of the weight describes that the variable must be reflected more importantly than others. In this research, these weights were manipulated by the analytical hierarchy process (AHP) method (see Table 6.2). In the next part, it will be depicted how this technique was employed.

6.5.1.2 Implementation of the AHP Technique

The analytic hierarchy process is a well-known methodology for specifying appropriate weights by cross-comparing all factors against each other with reproducible preference factors. This basis of this model works principally on expert knowledge and preferences values. The mentioned variables in Table 6.1 were crossed and compared in paired units. Preferred weights were input according to the expert knowledge and similar studies, and the gained weights for the variables were imported to Table 6.2. This table reveals the resultant weights after satisfying the consistency ratio. Verification of the proper consistency ratio was obligatory, to ensure this implementation. In fact, it was needed to iterate this technique until the optimal consistency ratio was obtained. This value should not exceed 0.1, i.e. lower values verify the accuracy of weighting.

By means of implementing the weighting system as well as applying the designed formula (6.1) in the ArcGIS Model Builder (Fig. 6.1), a categorical probability surface was produced. This surface identifies the preferable locations for settlement by the resident agents. Pixels with higher values indicate more probable sites for settlement of the resident agents based on their preferences. Figure 6.2 is the categorical probability surface that has been produced by resident agents.

6.5.2 Developer Agents

In the study area (i.e. Tehran metropolitan area), developers are the key component of the agent-based model and play a significant role in influencing residential development. The developer agents consider the preferences of residents in home buying as well as the governmental policies in land resources supervision. In other words, the developer agents can be affected by resident agents' preferences and government agents' restrictions, which can be interpreted as the interaction between these agents. This interaction influences the developer agents, who are interested in maximising profit, as to where to invest their funds. This means that resident agents prefer to select their settlement properties from the developers' property portfolio. Equally, the developer constructs buildings where they assume the resident agents prefer to live. The primary factor is to achieve a certain amount

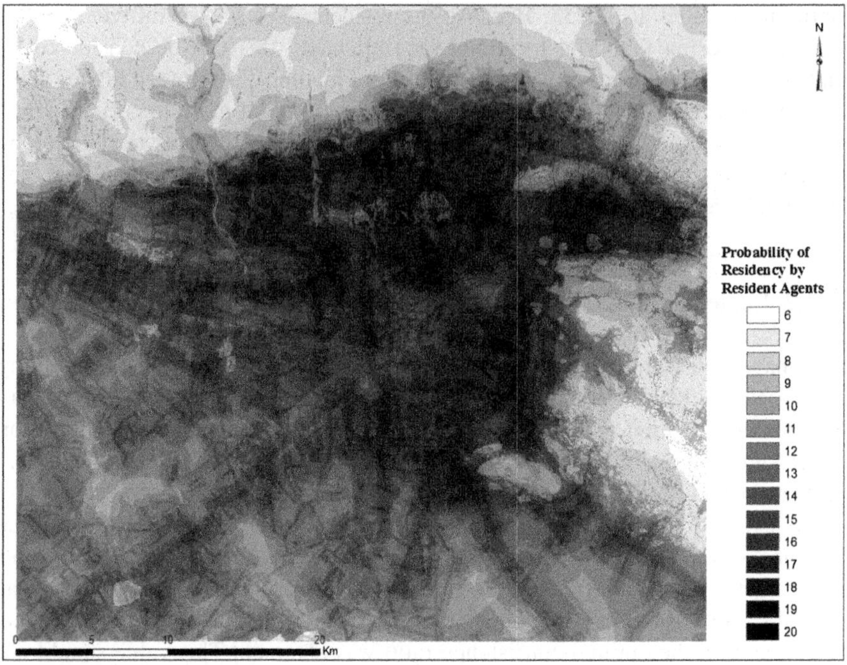

Fig. 6.2 Probability surface for settlement produced by the resident agents

of profit above market expectations. This criterion was used to conclude the decision behaviours of developer agents. Equation 6.2 was used in order to project potential development. Li and Liu (2007) developed this formula in their investigation in a Chinese case study, which estimates the investment profit of developer agents.

$$D^t_{\text{profit}} = H^t_{\text{price}} - L^t_{\text{price}} - D^t_{\text{cost}} \tag{6.2}$$

Where D^t_{profit} represents the investment profit, H^t_{price} is housing price, L^t_{price} is land price and D^t_{cost} is development cost. These prices were calculated in the domestic currency unit (i.e. Rial). Accordingly, the probability of development by the developer agents can thus be represented in the following equation (Li and Liu 2007):

$$P^t_{\text{developer}}(k, ij) = \frac{D^t_{\text{profit}} - D_{\text{tprofit}}}{D_{\text{mprofit}} - D_{\text{tprofit}}} \tag{6.3}$$

Where $P^t_{\text{developer}}(k, ij)$ represents the development probability related to the developer agents, D^t_{profit} is a threshold value and D_{mprofit} is the maximum value of the investment profit. Developer agents will invest in the site if the estimation is in favour of the development, according to Eq. 6.2. Therefore, the above mentioned

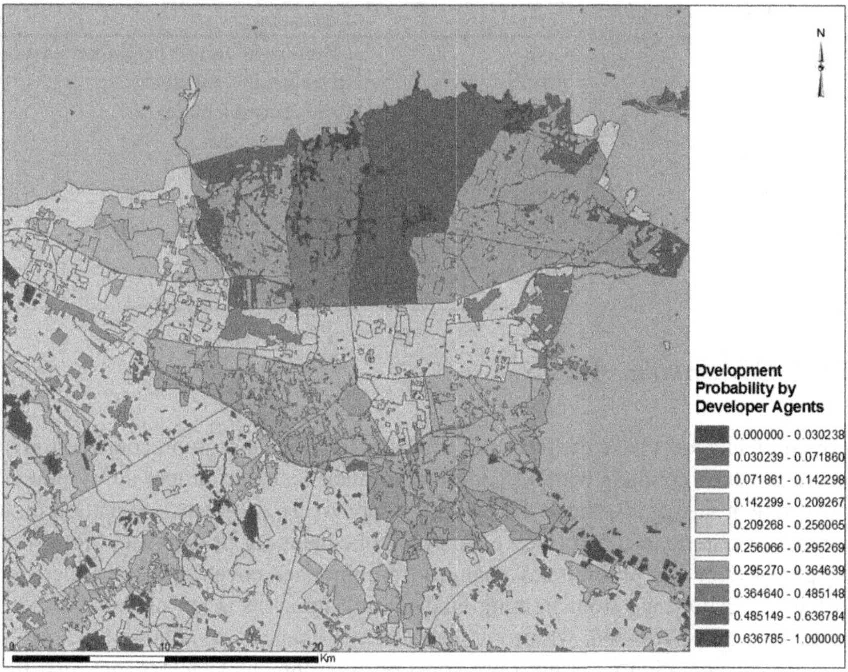

Fig. 6.3 Probability surface of potential development produced by the developer agents

formulas were calculated and input into the model. The land price dataset was produced by gathering data through field work and, additionally, this gathered data was integrated with other related reports from the national statistic centre.

Given these considerations, the developer agents would prefer to develop constructions according to Fig. 6.3. Changes will be applied after approving the application through the government agents; thus, government agents play an important role in this process. More recently some governmental restrictions have been approved to stop new constructions in the vicinity of Tehran. Governmental policies can also act in order to favour both the developer and resident agents' decisions, or act against their decision. In fact, whereas the prices of farming lands and open lands are extremely lower than regular land prices, these agents, nevertheless, prefer to develop land in those cheaper regions, which is more affordable for both agents (resident and developer). However, the government agent is ultimately empowered to reject any building applications where the construction might destroy open lands and green spaces. That is why it was necessary to define a specific agent, so-called government agent.

However, other governmental policies favour land change; for instance, Karaj (the biggest city in the metropolitan area after Tehran) was recently promoted from a city to a province. Whilst this action was taken to shift a portion of the population to this province, the potential exists for Karaj to become another mega-city like Tehran.

Table 6.3 Denotation of the terms in government agents' formulation

Term	Denotation
GAB	Government agents' behaviour function
RSRZ	River streams risk zone
RNB	Roads network buffer
HB	Highways buffer
ASB	Airports risk buffer
MFRZ	Military facilities risk zone
PFRZ	Power facilities risk zone
PB	Parks buffer
NSS	Non suitable slope

6.5.3 Government Agents

Government agents have the right and authority to allow or prohibit construction in any area which conflicts with, or conforms to, their vision of proper development. Government agents also have the power to approve or deny any application on a number of criteria, and can reserve land in public areas for its own use. Moreover, the government can prohibit any changes from one land cover type to another; for instance, land change from farming land to built-up areas is not allowed without governmental permission, which can carry heavy financial penalties.

This consideration does not only concern environmental factors, but also the interaction between the resident and developer agents. Government agents consider the suitability of any residency and construction, based on the current land use situation, surrounding environment, transportation supplies, affordable general facilities, and educational benefits. Existing land use is a key factor in determining land use conversion. By definition, different land uses have different purposes and possibilities for land conversion. For instance, no land change is permitted in steep areas, or within permitted parameters of waterways.

The possibility for land development in water bodies or mountainous areas is extremely low. Besides, the probability also depends on the pre-planned development plans. Thus, an application will only be approved providing no conflict exist with present planned land usage. The behaviours of government agents can also be affected by any unexpected behaviours of resident agents and developer agents (e.g. high migration rate, unexpected population growth, new interchange expansion, inordinate development applications, due to new policy approval). On the other hand, government agents must examine the attitude of residents in terms of affordable places to live, based on their preferences. Although, the pre-planned map could be entered as the input file, unfortunately accessibility to this type of data is limited. Hence, it was necessary to achieve it by expert inferences. According to the above explanation, the following function can be provided.

$$GAB = F\,(RSRZ,\ RNB,\ HB,\ ASB,\ MFRZ,\ PFRZ,\ PB,\ NSS) \qquad (6.4)$$

Significantly, government agents' behaviour is a function restricted by the components in Table 6.3. The terms mentioned (see Table 6.3) describe the denotations

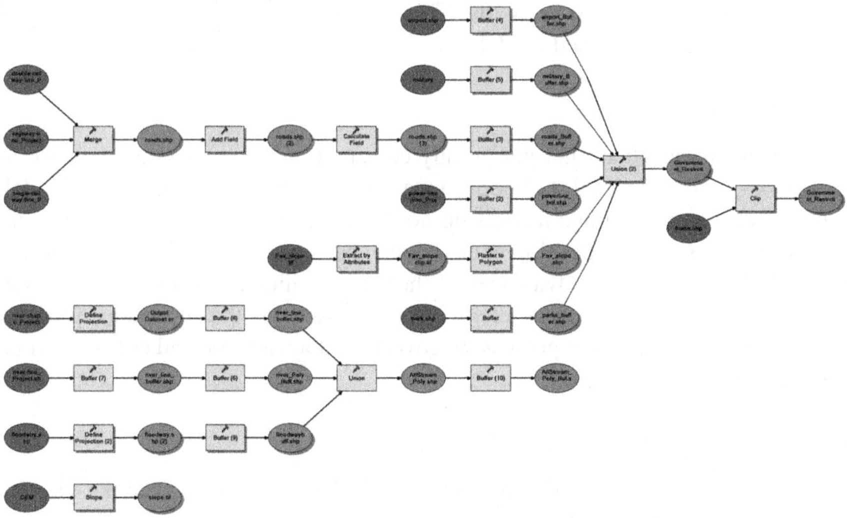

Fig. 6.4 Schematic view of the designed geospatial model for the government agent

of the above function components, where government agents are supposed to apply governmental restrictions for development. Also, there are some risk zones that are identified to avoid any development due to probable natural risks. Therefore, the prohibited areas by these agents were discovered by expert knowledge.

This agent was geospatially modelled by means of ArcGIS model builder. These variables have to be taken into consideration as binary type. The schematic representation of the designed model is shown in Fig. 6.4.

- River Streams Risk Zone

This variable allows the government agent to restrict new construction in the vicinity of rivers and streams. In order to take this component into account, a certain buffer distance was implemented to isolate the risk zone from other unlimited areas. Therefore, any construction application which is located in this risk zone will be refused by the government agents. Other places outside this area will be considered safe for construction.

- Roads Network Buffer

It is not allowed to construct any building at a certain buffer distance which varies depending on the type of road network. A certain buffer distance permanently exists to forbid any construction in the surrounding area of any road networks. Thus, the government agent has the authority to prevent any building construction within this particular distance. For each type of roads, different thresholds need to be taken into account. For example, no development is allowed to take place within 30 m of roads. This assumption was executed as another component of the mentioned model.

- Highways Buffer

Government agents do not permit any construction aside a certain distance to the highways, therefore, this buffer distance was considered in the designed model. Some other land use change models do not take account of this tiny variable and allocate these sites for development, which can decrease the accuracy of the results. This is one of the advantages of the ABM, which takes into consideration the smallest components and increases the accuracy of the modelling process. During the decision-making process by government agents, no land change will be permitted in the study area.

- Airports Risk Buffer

Government agents do not allow any non-affiliated construction in the vicinity of airports, which has to take place at a specific distance away from air terminals. The study area has some public and non-public airports, and this component was imported into the designed model.

- Military Facilities Risk Zone

The area of Tehran includes several military garrisons and organisations that have to be set apart from residential construction. Settlement in the proximity of these military garrisons is not safe for residents and the government agent cannot allow any construction within these perimeters. Therefore, the developer agents have to explore other suitable areas for development.

- Power Facilities Risk Zone

Power facilities, such as high voltage transmission lines and other power instruments, are considered public risk zones and are off limits for residency. Because settlement in the proximity of these facilities is dangerous, these areas were excluded from non-risk places in the model.

- Parks Buffer

The government agent cannot allow the developer agent to construct new housing near parkland. Because parks provide clean air for local residents and allow for leisure, any construction near parks is forbidden. The government agent will, therefore, consider a buffer distance from this protected area. This component of the government agent was also carried out in the designed model.

- Non Suitable Slope

Tehran is surrounded by high and steep mountains in the northern and eastern part of the city. Therefore, the topography of the study area causes serious concern

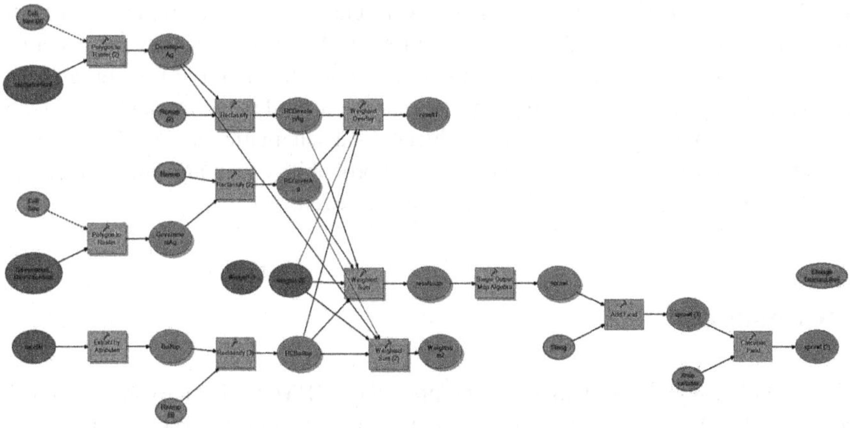

Fig. 6.5 Schematic view of the designed model for the agent combination process

for development and has to be considered for this simulation. Government agents will not allow construction in steep areas, and because low steep locations can be used for mixed uses and farming, a high threshold value was implemented in the model.

6.5.4 The Agent Combination Process

Three agents were provided in the study model, namely developer agents, resident agents and government agents; also the potential behaviours of those agents were imported into the model.

Each agent type was modelled separately to consider possible behaviours and actions. For instance, possible variables which are capable of influencing the resident agents, and thereafter a potential probability surface for this agent type were produced. According to the previous definition of agents, agents have internal interaction between themselves as well as external interaction between other existing agents.

Thus, in this research the interaction between agents had to be considered in the agent combination process. This would allow us to reach the appropriate results. The possible interaction between agents was taken into account, besides which the change demand factor was also another variable which controls the final outputs. In other words, the agent combination will result in a potential map of change that has to be coordinated with the change demand. Change demand was already calculated by two different techniques which were discussed previously (see Sect. 4.10).

In fact, two different change demand estimation methods were applied: The Markov model and statistical extrapolation. Therefore, we could estimate the quantity of impending changes according to these two different scenarios. These two quantification methods were applied to obtain predicted land use maps of 2016 and 2026. The potential cells for change were chosen and converted according to the change demand through a cellular automata function. The CA code begins allocation of changes from the nearest neighbours in the vicinity of the urban area. The model does not stop running until it reaches a predefined amount of change, and then the final change map was produced. The schematic view of the designed model is represented in Fig. 6.5. The resulted probability map and simulated maps of 2016 and 2026 will be presented in chapter seven.

6.6 Summary

The methodology for implementing the presumed ABM has been explained in this chapter. The assumed agents (i.e. resident agents, developer agents, government agents) were created and their preferences and behaviours projected. The spatial distribution of those agents was explained, and a weighting system was applied in the model to differentiate between the effective variables in resident agents. Finally, a probability surface which demonstrates the potential cells for development was produced. This probability surface enables us to allocate future changes based on appropriate scenarios for change demand determination. The possible scenarios to estimate the quantity of change has already been discussed (i.e. statistical extrapolation, Markov chain prediction). Two scenarios for future development were employed and two different simulated maps were produced. These maps and scenarios will be depicted in the next chapter.

Reference

Li X, Liu X (2007) Defining agents' behaviors to simulate complex residential development using multicriteria evaluation. J Environ Manag 85(4):1063–1075

Chapter 7
Analysis of Results

7.1 Introduction

In this chapter, it is intended to present the obtained results arising from different approaches, from which these results and outcomes will be analysed and discussed. The discussion will start by analysing all achieved results beginning with the spatio-temporal analysis of change, and thereafter, each particular model with the assigned methodology will be depicted. The traditional and recent methodologies in land use change studies will be challenged.

7.2 Data Gathering and Management

It was crucial to check the accuracy and scale of the employed data in this research in order to assure whether such data were valid or not. As it was already mentioned in Chaps. 3 and 4, the used data were categorised into two main categories:

- Gathered data through national geodatabase and domestic organisations.
- Retrieved data through satellite imagery and remote sensing approaches (e.g. land use maps extraction and correction) (see Chap. 4)

The accuracy of the national geodatabase was already approved by national organisations, and the data were produced at scales 1:2,000, 1:25,000, and 1:50,000. Moreover, the temporal socio-economic data such as land price, demography data were produced by the Iranian Statistic Centre. Additionally, Landsat imagery was used to correct the prepared land use maps of 1986, 1996, and 2006. Consequently, overall accuracy of 0.91, 0.88, and 0.90 were obtained for 1986, 1996 and 2006 maps, respectively. The Kappa Index adjusts the fraction of appropriately categorised cells by subtracting the estimated contribution of chance agreement. As a result, the obtained values of the Kappa Index (91, 88, 90%, respectively) are significantly

J. Jokar Arsanjani, *Dynamic Land-Use/Cover Change Simulation: Geosimulation and Multi Agent-Based Modelling*, Springer Theses,
DOI: 10.1007/978-3-642-23705-8_7,

better than random classification. Thus, the higher values return better and more accurate results. Hence, the accuracy of extracted land use maps of 1986, 1996 and 2006 were accepted, and given this assurance these maps could be employed in the modelling process. All gathered data were stored in a local geodatabase for further tasks and analysis.

7.3 Spatio-Temporal Change Mapping

Land use maps of 1986, 1996 and 2006 were arranged into five main categories as previously mentioned [i.e. agricultural fields, built-up area, open land, public parks and water bodies (see Sect. 4.4)]. The basic structure of the metropolitan area of Tehran shows a centric pattern, surrounded by open lands in north and east, and agriculture fields in the southern and western parts of the area. Hence, if any change is going to take place it has to occur in open lands which are preserved generally by the government, whereas agricultural fields are preserved by the inhabitants and farmers.

Landowners prefer to convert their farming lands into built-up areas, or their low height buildings to higher elevated buildings in order to own more apartments. In this way it is possible for them to benefit financially. Temporal mapping and change trend analysis prove this fact, although it must be restated that governmental authority constrains construction and carries financial penalties when laws are breached.

According to Table 4.2, an obvious trend of expanding the built-up areas on agricultural lands and open lands has taken place. These changes reveal a trend in favour of increased built-up area expansion, from 24 to 28% between 1986 and 1996 to 32% between 1996 and 2006. This can be interpreted as an 8% increase within 20 years. Conversely we see a reduction in the farming areas from 24 to 22% between 1986 and 1996 and then to 21% between 1996 and 2006. Altogether this means that a 3% change has taken place in the total area within 20 years. Moreover, a decrease over the open lands can be observed from 50 to 47% between 1986 and 1996 and to 44% between 1996 and 2006.

This obvious and rapid urban growth can be simply defined as urban sprawl. This is manifested amongst residents and housing developers in the increased availability of very affordable high-rise properties. Therefore, urban growth is being channelled in two main directions, through vertical and horizontal expansion. Whilst vertical growth is not the concern of this research, it can be interpreted as a high rate of population growth.

7.4 Analysis of Socio-Demographic Changes

Moving away from the causes of vast migration to the Tehran metropolis that was described in the previous chapters, it is intended to analyse the trend of changes (from a socio-demographic point of view) which have occurred within this 20-year period. The socio-demographic changes between 1986 and 2006 can be a good lead to track forward change. As mentioned in the previous chapter (see Table 4.2), there are some significant factors to consider:

- The population of Tehran has been increasing rapidly since 1966, with the likelihood of it continuing to grow.
- The population share of Tehran metropolis, within Tehran province as a whole, has been decreasing noticeably, which can be interpreted that the majority of inhabitants prefer to settle in suburban areas.
- The growth rate of nearby cities, within this period, is even higher than Tehran.
- All these factors can alter irrevocably the complexity of the metropolis, thus causing numerous complications and difficulties for urban managers and planners.

7.5 Findings Through the Traditional LUCC Modelling Approaches

Four main well-known methodologies over land use/cover change have been carried out in this research to make a better comparison between existing approaches. Having results from different methodologies affords more information to compare these methodologies and find out their advantages, disadvantages, weaknesses and strengths. As mentioned previously, the traditional approaches such as CA, Markov chain, CA Markov, logistic regression and ABM were executed. In this section, the obtained results will be brought to discussion. Thus, we will start with the cellular automata approach.

7.5.1 Cellular Automata Scenario Results

Following Fig. 5.3, each step was done according to the flowchart, predefined values and model validation process into two main steps (e.g. training phase and simulation phase). The training phase was carried out to compare its outcomes under different transition rules. The training phase was a useful way to verify the model validation process. Since different transition rules and kernel sizes caused different outcomes, it was vital to check the obtained results with reality and apply the optimum characteristics to the final model. Such a tested model can help to predict forthcoming changes more accurately, and a cross comparison of kernel sizes and transition rules (e.g. either Von Neumann or Moore) can be considered as a part of transition rules. Among 3×3, 5×5 and 7×7 kernel sizes, kernel size 3×3 acted more accurately than others, due to its small size and repeating more iteration. Transition rules definition was another important measurement of the

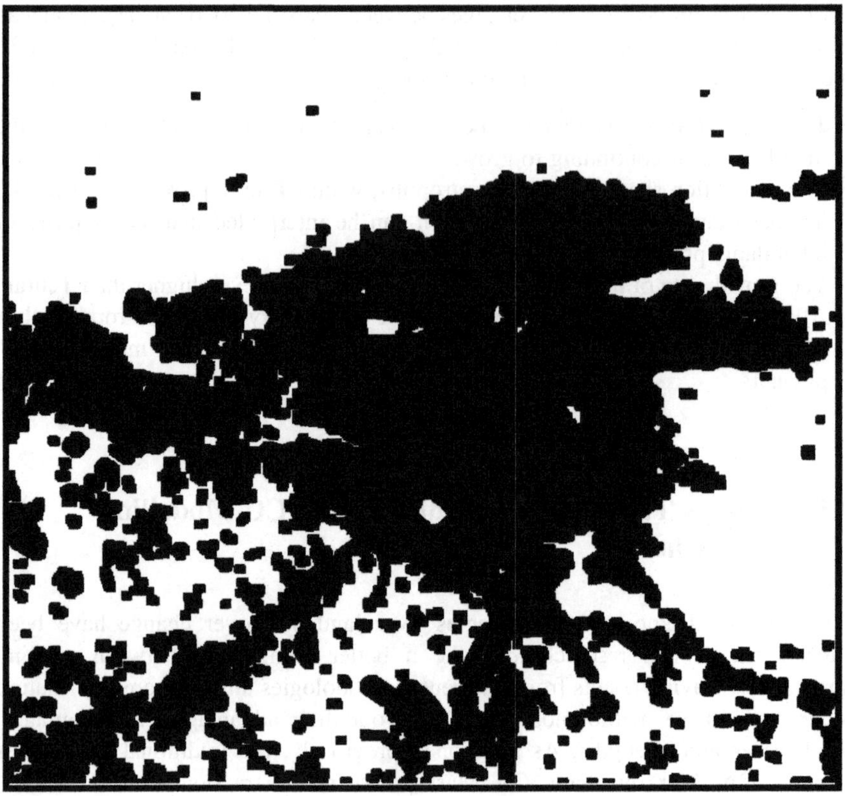

Fig. 7.1 Simulated built-up map by the optimal CA approach

model and the results show that the Von Neumann model had closer results to the reality. In this regular CA model, no other environmental and economic variables were input and this approach was designed only by defining a variety of transition rules. Figure 7.1 presents the predicted map of the built-up area for a 10-year period, which is 2016.

7.5.2 *Validation of the CA Approach*

Certainty of a model is basically needed when the accuracy of outputs is required. Certainty of the implemented cellular automata model is typically required when it was employed to simulate land change. It was aimed to implement this model under different circumstances in terms of transition rules and the number of iterations, in order to find out which situation could return more accurate outcomes.

The training phase, which is an evaluation process of achieved outcomes, is an assessment procedure for determining the accuracy of the results and the model.

Different indices could be used to reach this aim. ROC value was an important index for this part of the research. The other possible way, according to Li and Liu (2007), is to compare the simulation patterns between this approach and other produced results, which will be acted upon at the end of this project. ROC has this capability to compare maps in two main ways: geospatial and quantity measurement. Automatic calibration techniques have been used to help develop the exploration for appropriate parameter values of cellular automata model. First, the simulated maps of 1996, 2006 at different iterations were picked for comparison with the actual maps of 1996 and 2006. The ROC method acts on different threshold quantities in a cell-by-cell overlay (see Chap. 2).

The produced ROC values were compared and the maximum value that satisfied the model was chosen for each period. Then, the optimum settings (i.e. iteration number and transition rules) were implemented based on the predesigned transition rules of the 2006 map to reach a simulated map of 2016. In fact, a visual interpretation was helpful to make a comparison between the outcomes. Nevertheless, these procedures encounter difficulties in resolving the calibration problems of CA. The designed CA model was simplistic, which did not take into account any bio-physical and socio-economic parameters; however, it could be integrated with those parameters to yield more accurate results (e.g. SLUETH; CLUEs models). In the next section, the results of the Markov chain model will be presented.

7.5.3 *Outcomes of the Markov Chain Model*

As mentioned in Chap. 5, the Markov chain theory is a stochastic process theory which describes how likely one state is to change to another state. In fact this module is not spatially explicit; therefore, it is not able to produce geospatial outcomes. Nonetheless, it can provide and calculate the amount of future changes based on the previous changes. As the process of change in the study area has emphasis on built-up areas, this module can predict the expected quantity of change, as well as the probability of the conversion of each particular cell to other existing categories. Therefore, the results arising from this model could be used for change allocation process.

The Markov chain module inputs a pair of land use maps and outputs a transition probability matrix, a transition areas matrix, as well as a set of conditional probability images (Oluseyi 2006). Some studies have integrated this module with other geospatial functions to create geospatial results (e.g. Mousivand et al. 2007). These matrices were calculated for the years 2016 and 2026. Since Markov chain has no spatially explicit behaviour to create spatial outputs, only transition area matrices were stored to use as a quantity estimator for the allocation process. Table 7.1 demonstrates the predicted quantity of change for 2016 and 2026.

Table 7.1 Quantification of changes for 2016 and 2026 produced by the Markov chain model in terms of hectare

		Agri lands	Built-up	Open lands	Public parks	Water body
2016	Agri lands	37,424	1,605	871	69	9
	Built-up	214	58,326	324	206	2
	Open lands	1,175	4,702	75,355	448	26
	Public parks	10	47	35	5,290	7
	Water body	4	2	0	1	158
2026	Agri lands	33,857	3,744	1,972	385	21
	Built-up	17	58,398	347	305	7
	Open lands	1,537	9,239	70,096	806	28
	Public parks	0	234	107	5,046	3
	Water body	2	0	0	0	162

7.5.4 The Markov Chain Model Validation

The Markov model is not a spatially explicit model to spatialize the location of changes; however, this model is able to quantify the amount of change within a specific period. Although it seems this model cannot afford significant results for spatial analysis, the outcome of this model can be utilised to allocate the changes by means of the geospatial-based functions (Mousivand et al. 2007). For instance, local analysis of change probability and change potential can help to localise the predicted amount. In this research, the outcome of this implementation will be combined with other approaches to improve the accuracy of other geospatial approaches.

7.5.5 Outcomes of Cellular Automata Markov

In the Cellular automata-Markov model, the Markov chain process manages temporal dynamics among the land use/cover categories based on transition probabilities, while the spatial dynamics are controlled by local rules determined either by the cellular automata spatial filter or transition potential maps (Maguire et al. 2005). This approach was carried out according to the designed flowchart (see Fig. 5.8), thereafter, land use maps of 2016 and 2026 were simulated. This model benefits from a combination of the Markov chain model and the cellular automata model to spatialize the estimated amount of change from the Markov chain model. Moreover, this module is able to input multi-class maps, which is useful to simulate conversion between the other existing categories (i.e. agricultural lands to public parks, open lands to agricultural lands, etc.). Thus, the final map can be a categorical map of the existing land use classes.

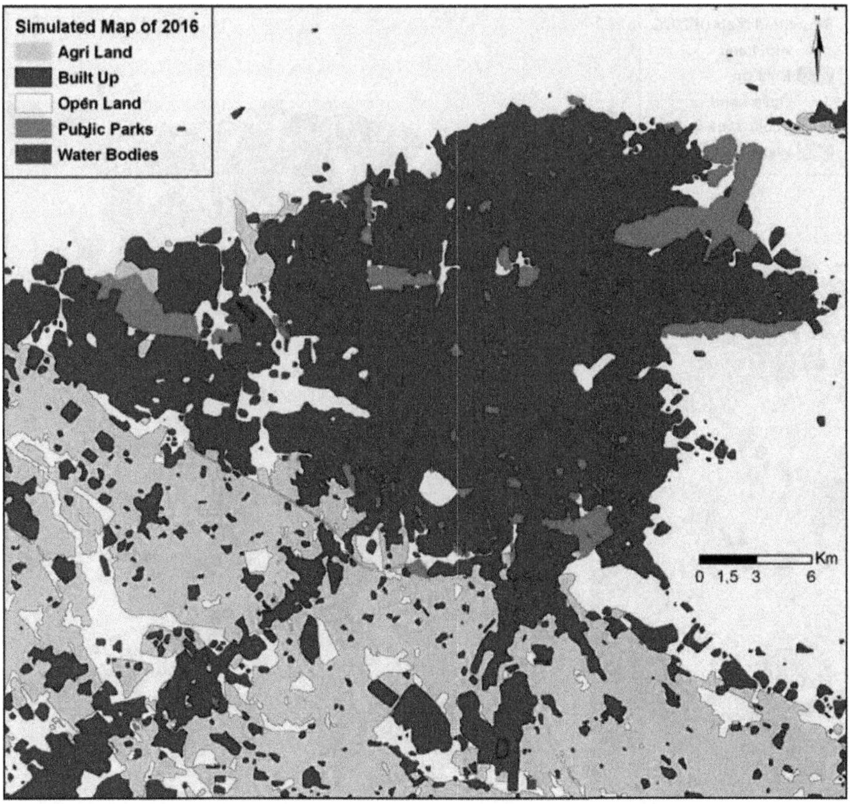

Fig. 7.2 Simulated land use map of 2016 with the calibrated CA-Markov model

Consequently, validation of the simulated map and the actual land use map of 2006 was employed to evaluate the model. Therefore, for validation of cellular automata Markov results, in terms of location and quantity, a diagram of correlation between different simulated land use maps of 2006 and number of iterations was drawn. Thus, the Kappa indices of location and quantity were calculated separately and, subsequently, the most appropriate iteration number at 300 iterations was determined. A Kappa standard index at 0.91 was picked as the optimum point, which shows a close correlation between the simulated map and the actual map. As a result, the predefined rules were considered for running the module for prediction procedure (see Fig. 5.10).

Whereas the training phase of this approach was tested successfully, the calibrated model was utilised to simulate land use maps for the next 10 and 20 years (i.e. 2016 and 2026). This was done by means of the produced transition probabilities matrices and Markovian conditional probability images. As a result, the simulated land use maps of 2016 and 2026 were produced, which are shown in Figs. 7.2 and 7.3.

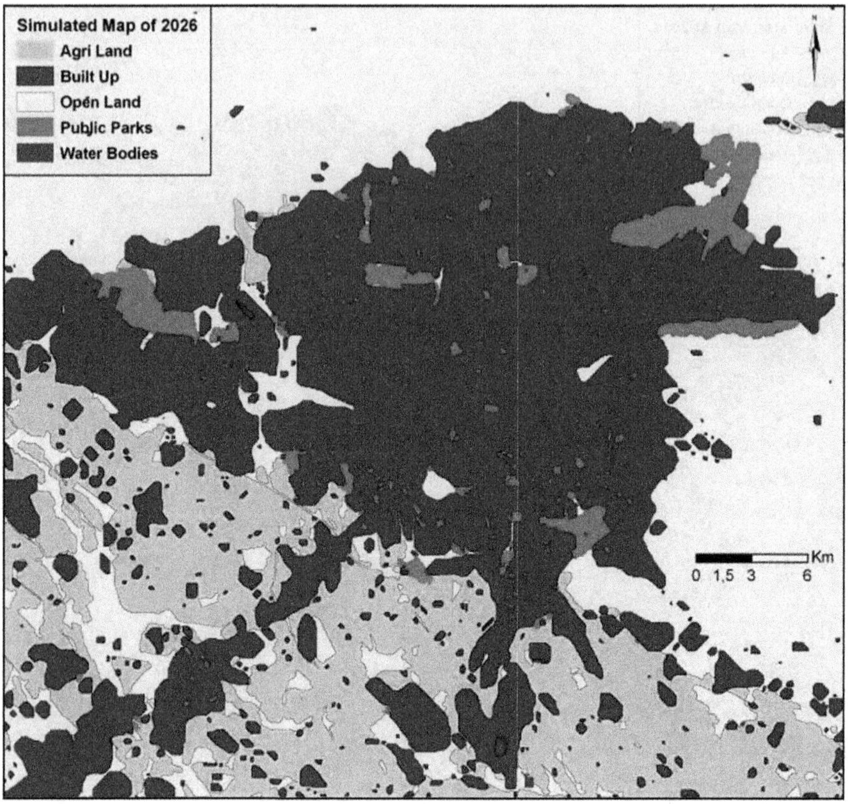

Fig. 7.3 Simulated land use map of 2026 with the calibrated CA-Markov model

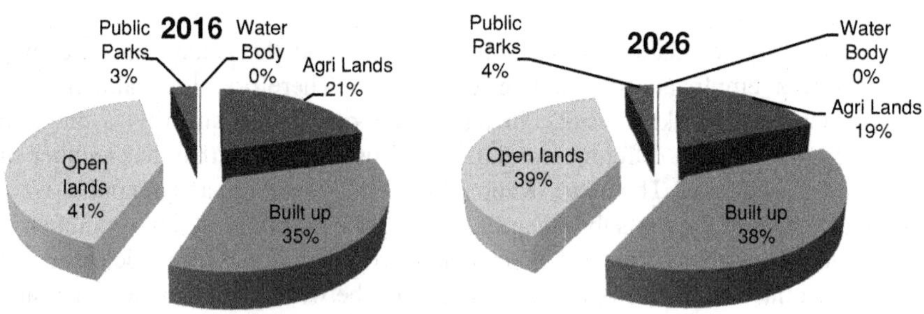

Fig. 7.4 Predicted percentage of each land category for 2016 and 2026

According to the simulated land use maps, the situation and percentage of each type of land use map is illustrated in Fig. 7.4, which demonstrates a 3% increase of built-up area (i.e. 5,567 ha) and another 3% decrease in open land area. Moreover, for 2026 it has been predicted there will be a 3% increase of built-up area and a 2% decrease in open land area.

7.5.6 Validation of the Cellular Automata Markov Model

The CA Markov model utilised the neighbourhood rule to simulate the conversion of a land use/cover class near the existing similar land use/cover class (Pontius and Malanson 2005). For validation of this model, a training phase was accomplished to compare the output maps with the actual land use map of 2006. The highest value of overall accuracy at 91% was reached (see Fig. 5.10). Analysis of the simulated land use map of 2006 reveals that the cellular automata Markov model, generally, is an excellent estimator in terms of change quantification and continuous-space modelling. However, the visual interoperation reveals that some diffused-speckle pixels can be observed. In reality, these speckle noises do not exist. Furthermore, this model needs to run several times to provide the most accurate results.

7.5.7 Outcomes of the Logistic Regression Model

A set of independent variables was adjusted after statistical assessment of all environmental and economic variables. As described previously, the change during a 10-year period was the dependent variable. Several sets of independent variables were imported to the logistic regression model in order to calibrate this model by itself, with the support of IDRISI Andes GIS software. A mask over all input data was employed to create equal dimension raster files. Then the prior produced land use maps of 1986, 1996 and 2006 were used to specify the change maps within the 1986–1996, 1996–2006 and 1986–2006 cycles. The statistical evaluation of the retrieved ROC values and adjusted odd ratios for each set of combined variables (see Table 5.3) were appreciable tools to determine the best predictor variables.

Thus, a calibration process was carried out to retrieve the predictor variables. All available and affordable data were classified into three main classes (i.e. social variables, econometric and biophysical variables). The existence of spatial correlations was also checked. Model calibration in this study was made as two steps including initial calibration and refining, respectively. All required data were converted to raster format at 30 m resolution. In order to pick the optimal variable set, it has to reach the highest ROC value. In fact ROC $= 1$ indicates a perfect fit, while ROC $= 0.5$ indicates a random fit and values in between have a degree of membership. Besides, a higher adjusted odds ratio is necessary for a better fit and greater confidence. The optimum set of variables was chosen. The highest ROC value at 0.9532 allowed us to pick the appropriate set of variables as the input file, which was shown in Table 5.2. A predicted change probability surface map and a residual map, indicating the difference between predicted and observed probability, were generated.

Therefore, in order to allocate the proper quantity of change according to the probability surface, two choices could be chosen: either to select the transition area

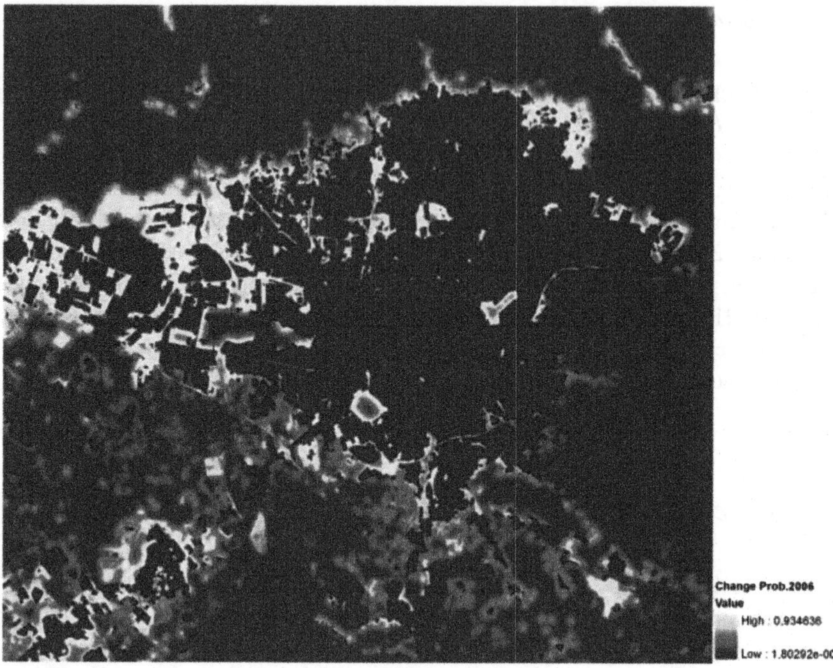

Fig. 7.5 Transition surface map of the study area produced by the logistic regression model for 2006 onward

matrix obtained by the Markov chain model, or the predicted population and construction per capita index. The transition matrix arising from the Markov chain result was picked to quantify the amount of change. Figure 7.5 demonstrates the produced probability surface that identifies the probability of change for each particular cell. Obviously, this model has ranked highly the nearest cells to the developed cells.

Based on the probability surface, as well as the change demand quantity, the allocation function was exerted on the transition surface map (i.e. Fig. 7.5), in order to produce the land use maps of 2016 and 2026. This probability surface enables us to predict future changes at any time. It was intended to produce the simulated maps for the same years (i.e. 2016, 2026) for the final comparison, thus 2016, 2026 were selected. The simulated land use maps of 2016 and 2026 are represented in Figs. 7.6 and 7.7.

7.5.8 Validation of Logistic Regression Model

By means of this prepared probability surface, the quantity of change must be specified through possible techniques—either through the population growth estimation model or the Markov chain model. In other words, a footprint of

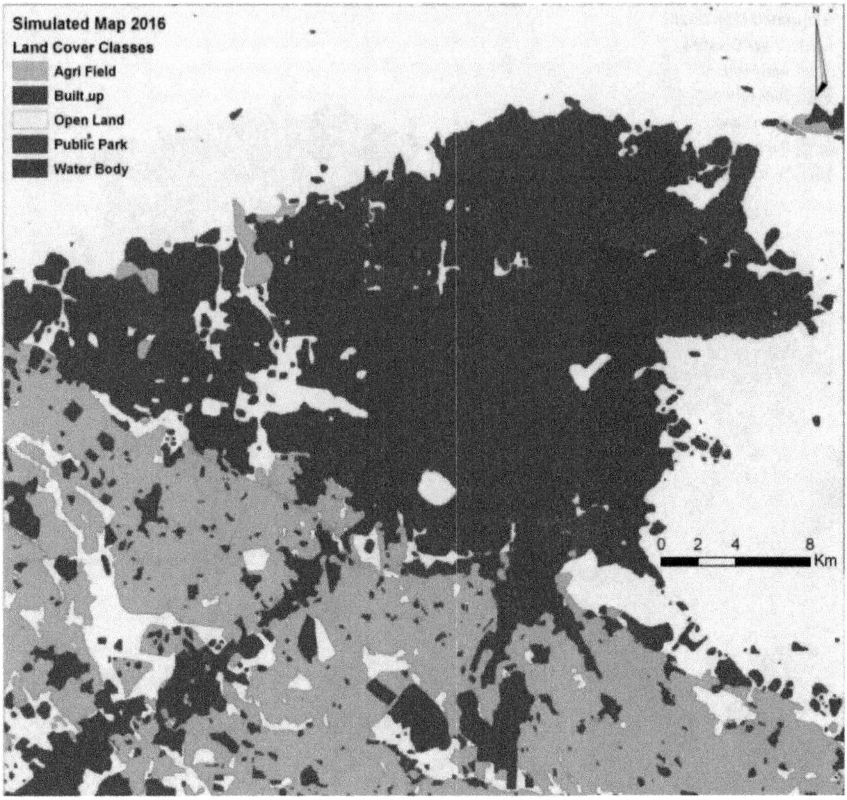

Fig. 7.6 Simulated land use map of 2016 with logistic regression model

population growth over the study area can be modelled through statistical methods and allocate this resulted quantity of change. The second method is to use a Markov chain model output, which is a matrix of change amount to allocate that amount as well as reach the predicted maps. The total amount of change was determined based on transition matrix of the Markov chain model to quantify the quantity of change to employ the allocation phase. Consequently, the implemented model was assessed by the model statistics. Thus, the validated model was employed to predict land use maps of forthcoming periods. In the next section, the outcomes of the geosimulation model will be depicted.

7.6 Outcomes of Multi-Agent Simulation

Three major agents (i.e. government agents, property developer agents, resident agents) were classified to associate the possible behaviours which drive the land change matter in the study area. All these agents were spatially and equally

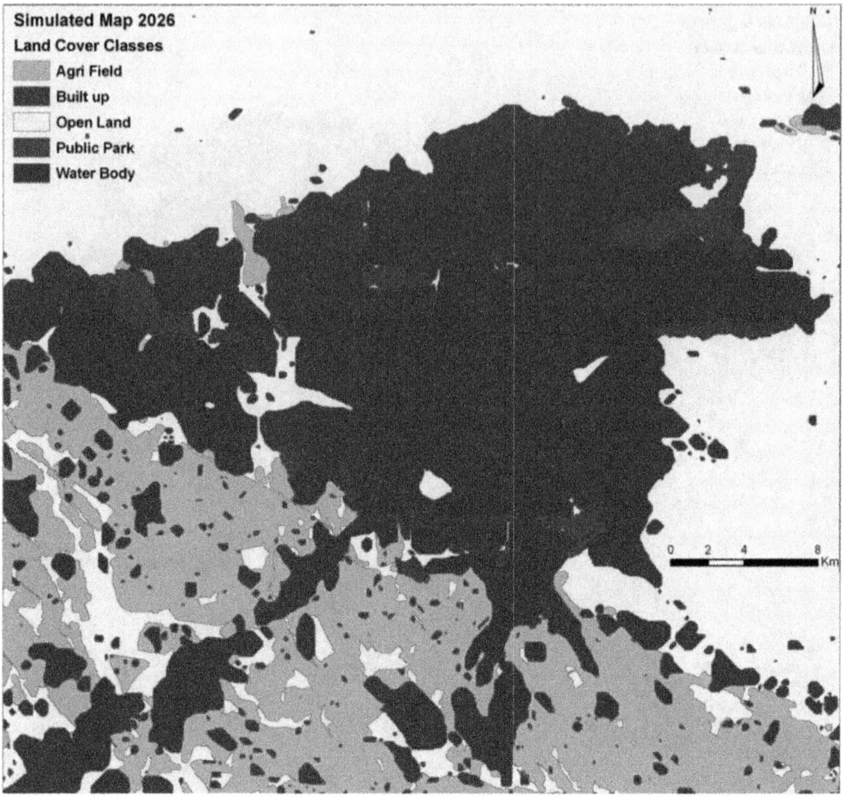

Fig. 7.7 Simulated land use map of 2026 with logistic regression model

distributed in the entire area (see Sect. 6.4) and have instant autonomy and major effects. Each predefined agent has its own specific behaviour and influence. In other words, developer agents are motivated to build new housings based on financial profit, whereas resident agents have preferences based on choice of area and lifestyle, according to multiple variables. It is significant that the developer agent has the influence to lead resident agents to choose their properties due to financial factors. More discussion of each particular agent will be depicted in the next sections.

7.6.1 Resident Agents

Resident agents make choices about where they want to live, based on desirability of location, and accessibility community services and communication, such as accessibility to the road networks, shopping malls, parks, leisure facilities, public transport, etc. It is obvious that if numerous buyers decide to select the same location, property prices will rise and the developer agents will construct new and intensive housing, resulting in neighbourhood overdevelopment. Thus, if the purchase and rent prices exceed their affordable threshold, the resident agent will have to look at other districts for accommodation. Therefore, property developers have to alter their investment policies and plans according to the residents' purchasing behaviour—their only objective, as a developer, is to earn as much profit as possible. However, developers must get the appropriate approval from the government before making changes to an area. In this sense the government is the final decision maker who has the power to approve the details of site development by considering environmental circumstances and internal policies.

Therefore, in considering all possible behaviours of resident agents, multiple factors were taken into account to maximise the efficiency of the model based on a designed utility function, to optimise the accuracy of the change allocation model. Detailed information about the utility function was presented in Sect. 6.5.1.

Fuzzification of existing factors was another approach that has been carried out to adopt the critical points of each factor in different ways. Thus, applying fuzzy membership functions seemed to be innovative for this purpose. Certain thresholds were applied for the proposed factors which were previously mentioned (See Sect. 6.5.1.1). The other important approach in this research was to weigh each variable based on its importance; therefore, a higher weight describes that the variable must be reflected more importantly than others. In this research, these weights were manipulated by AHP method. The AHP function was exerted to the entire factors in the ArcGIS environment. It was mandatory to ensure the value of the consistency ratio, because it should not exceed 0.1, which verifies the accuracy of weighting (see Table 6.2).

The extracted weights were issued in the overlay process in ArcGIS Model Builder. Therefore, a categorical probability surface was produced that identifies the preferable locations for residency by resident agents. The resulted probability surface map (see Fig. 6.2) demonstrates the most preferable area for settlement by resident agents with regard to environmental and economic situations and demographic circumstances.

7.6.2 Developer Agents

Developer agents are the key component of the agent based model and play a significant role in influencing residential development. The developer agents consider the preferences of residents in home buying as well as governmental restrictions in land development. In other words, the developer agents can be affected by resident agents' preferences and government agents' decisions. This means the interaction between these agents, with regard to the conduct of the

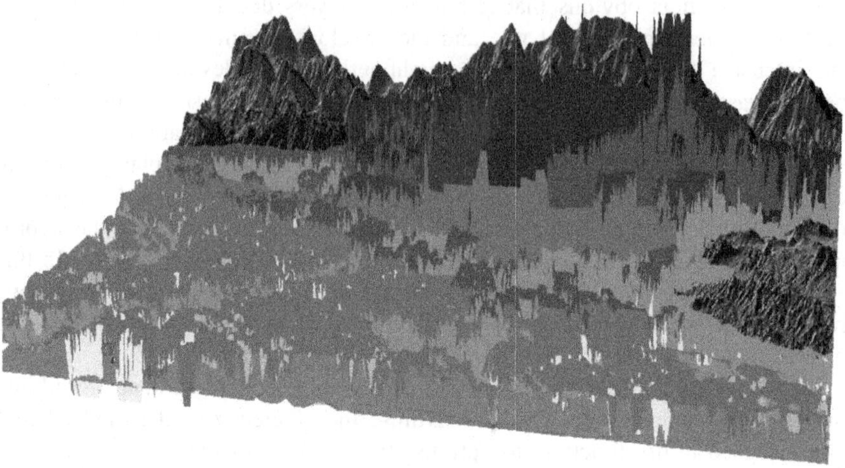

Fig. 7.8 Probability surface of potential development produced by the developer agents in 3D visualisation

developer agents on where to invest their resources in order to reach maximum benefits, is crucial. Therefore, this criterion was fundamental to the conclusions of the behaviours and decisions of the developer agents. Additionally, some equations were defined to reflect the development potential that was mentioned in the previous chapter. The economic viewpoint of decision making was modelled by importing a land price dataset. Some functions were exerted in the behaviour the defining process, which were explained in Sect. 6.5.2. A surface of development probability was produced by running the predesigned module, which is shown in Fig. 7.8.

Figure 7.8 represents the preferred sites for development in the study area by the developer agents. Obviously developers prefer to choose farming lands and open lands because of bigger profit potential; however, investment in high price sites carries a lower financial risk, since the price of land is already high and a return on investment is more assured. Moreover, many people want to settle in these areas due to neighbourhood facilities.

Eventually, it is the government agents who will apply their decisions on the received applications for construction and decide what is built and where.

7.6.3 Government Agents

Challenges between resident agents and developer agents are not the ultimate step of the final decision to be made for development. As mentioned, government agents are the final decision-makers who can either approve the development

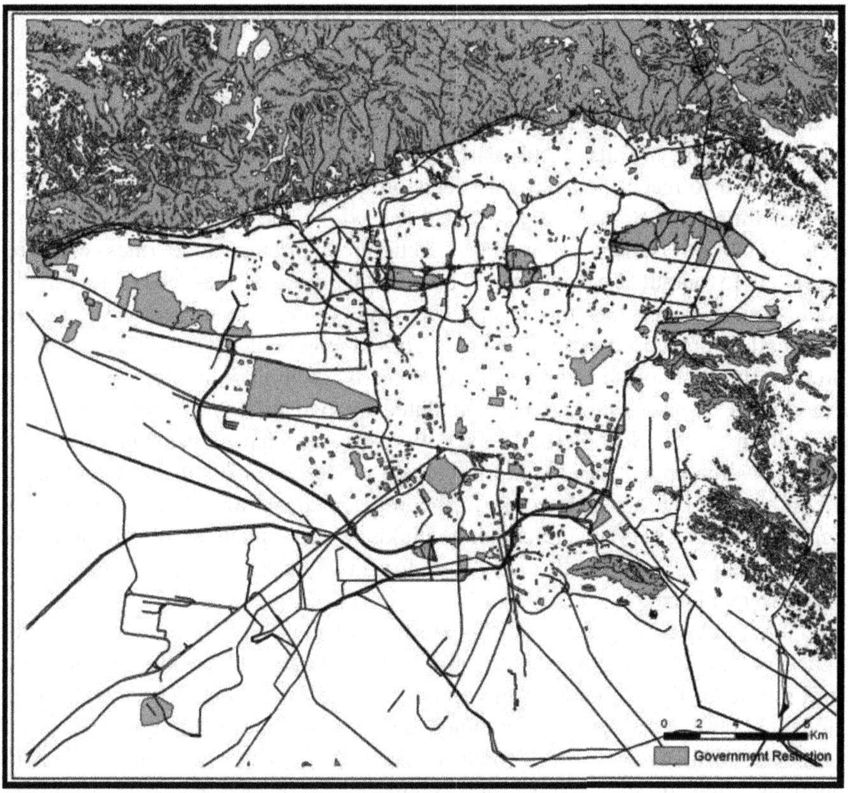

Fig. 7.9 Binary map produced by government agents

applications or reject them. This means that because the government agents have the final decision, based on unpredictable factors, mapping consistency of this issue can be problematic in the model. For example, this agent is not only concerned with environmental factors, but also by resident agents' and developer agents' interaction. The government agent also considers the suitability of any residency and construction based on the current land use status, surrounding environment, transportation access, affordable general facilities, and educational benefits. Apropos, a function was designed to yield the government agent output. In fact, the government agents' behaviour is a function restricted by those components mentioned in Sect. 6.5.3. A binary map produced by the government agents is shown in Fig. 7.9.

Figure 7.9 presents a binary surface that can be updated by latest policies issued by the government. This map demonstrates where any development application can be approved or rejected. It was not possible to gather any map from any source to indicate restricted places for development; however, we needed to take some variables into consideration to achieve this map.

7.6.4 Combination of the Agents and Their Interactions

Each agent makes its own decision and concludes its output, but the final outcome is an amalgam of all agent interaction and possible choices, and is central to obtaining accuracy. Thus, in this part the interaction between these agents had to be combined to reach their ultimate decision in land development.

Accordingly, some actions and decisions will come into conflict with other agents' choices. Hence, an overlay function comprising some rules was premeditated to output the decisions. The overlay function was coded to run the module until it reaches the change demand value. Consequently, the agent combination procedure produced a categorical potential map of change; therefore, we should allocate the probable changes by means of that potential surface. The change demand was already calculated by two different techniques, which were discussed in Sect. 4.10. In fact, two diverse change demand estimation methods were applied (i.e. the Markov chain model and statistical extrapolation). Therefore, we could estimate the forthcoming changes according to these two scenarios. These two quantification methods were applied to obtain predicted land use maps of 2016 and 2026. Therefore, two scenarios in this agent-based model were planned.

A CA function was coded to allocate the potential cells for change from the highest to the lowest, respectively. CA methodology begins change allocation from the nearest neighbours in the vicinity of the urban area. The model keeps allocating until it reaches a predefined amount of change; therefore, the final simulated change maps were produced. According to the two change demand functions, two different simulation scenarios were applied. The simulated land use maps of 2016 and 2026 through two mentioned scenarios are shown in Figs. 7.10, 7.11, 7.12, and 7.13. Thus, this combination follows the cellular automata technique and can be called CA-ABM.

7.7 Validation of the Simulations

Validation of this prediction and simulation model is basically required to assure the outcomes of model achievement. Validation of a simulation model could be carried out through a cross comparison between the simulated map and a map of reality. Obviously, no actual maps of 2016 and 2026 could be provided, therefore, this technique could not be applied. Thus, there are some issues regarding this validation process that will be noted below.

- Firstly, comparison of the simulated maps and the actual maps cannot be possible, because no maps of reality for 2016 and 2026 exist.
- Secondly, all the performed models met our satisfactions statistically (e.g. AHP), and also the calibration process of those processes verified that the ABM was being constructed perfectly at every stage.

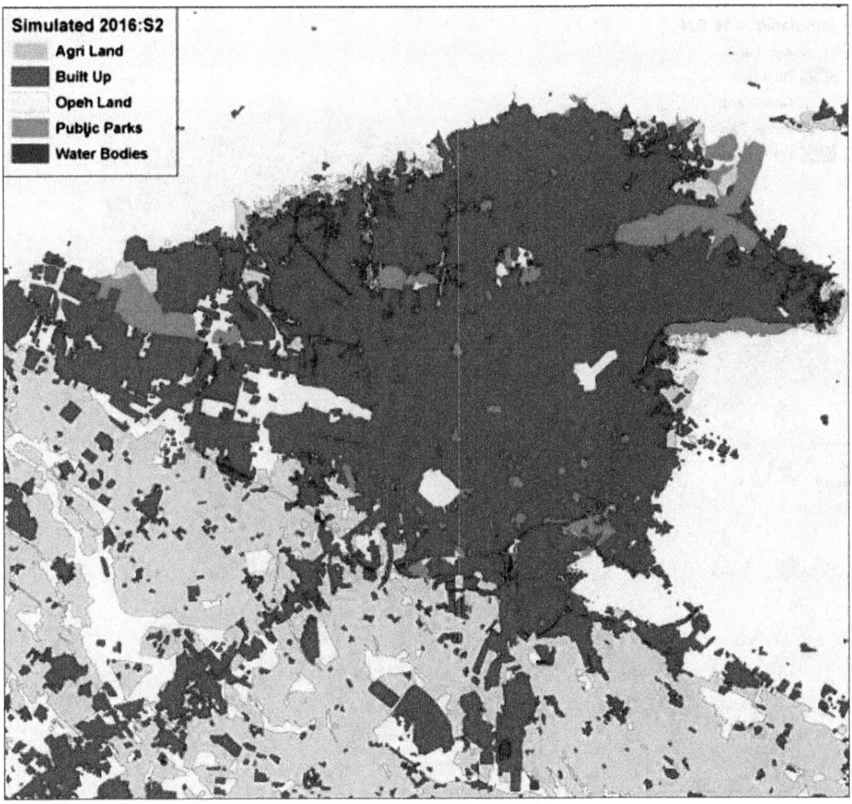

Fig. 7.10 Simulated map of 2016 produced by the designed ABM simulation based on scenario 1

- Finally, comparison of the simulated maps produced by the ABM method and other performed techniques was carried out visually and statistically. This will be depicted in the next section.

7.8 Comparison of the Employed Models

In the previous section, all the outputs from the implemented methods were presented one by one, i.e. cellular automata, the Markov chain model, cellular automata Markov, logistic regression and agent-based modelling. It is imperative to compare their results to achieve a conclusion in this research. A descriptive table of the approaches that were carried out is presented in Table 7.2.

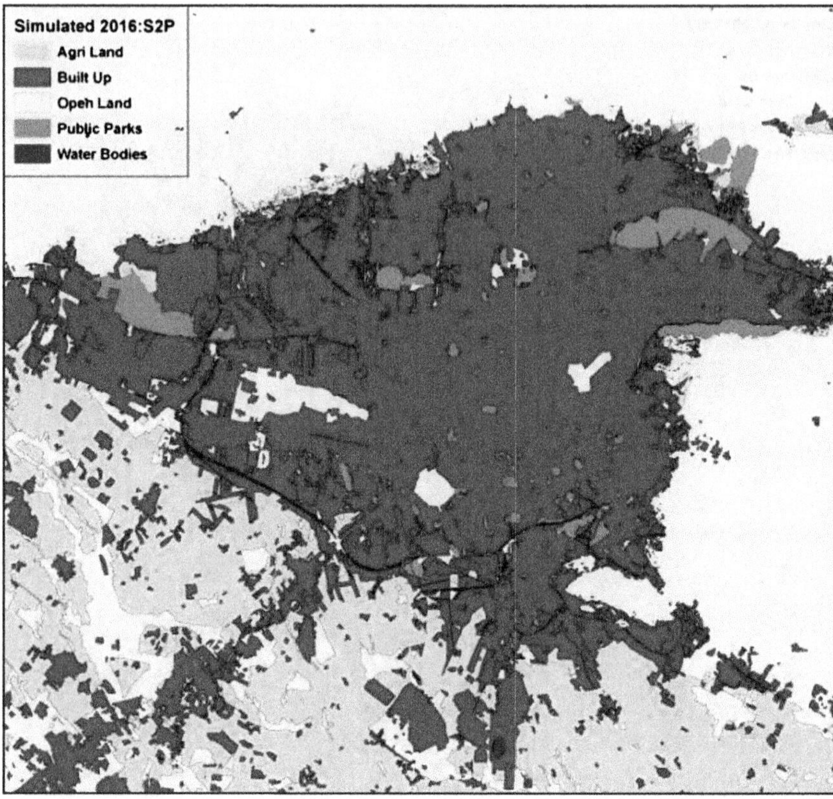

Fig. 7.11 Simulated map of 2016 produced by ABM simulation based on scenario 2

7.9 Discussion of the Outcomes

In recent years the world has been confronted with whole new challenges which have brought a severity of change on an unprecedented scale, such as climatic, economic and in technological revolution. The impact of globalisation has resulted in a huge shift in demographics, such as rapid urbanisation and intensive land-use conflicts in some developing countries. Therefore, simulation and forecast of urban growth is an important task for urban planners and landscape preservers to formulate sustainable development strategies. The simulation of built-up development can provide helpful and valuable information about future land demands and landscape changes. However, cities are complex systems that are complicated to characterise by means of mathematical equations (Li and Liu 2007).

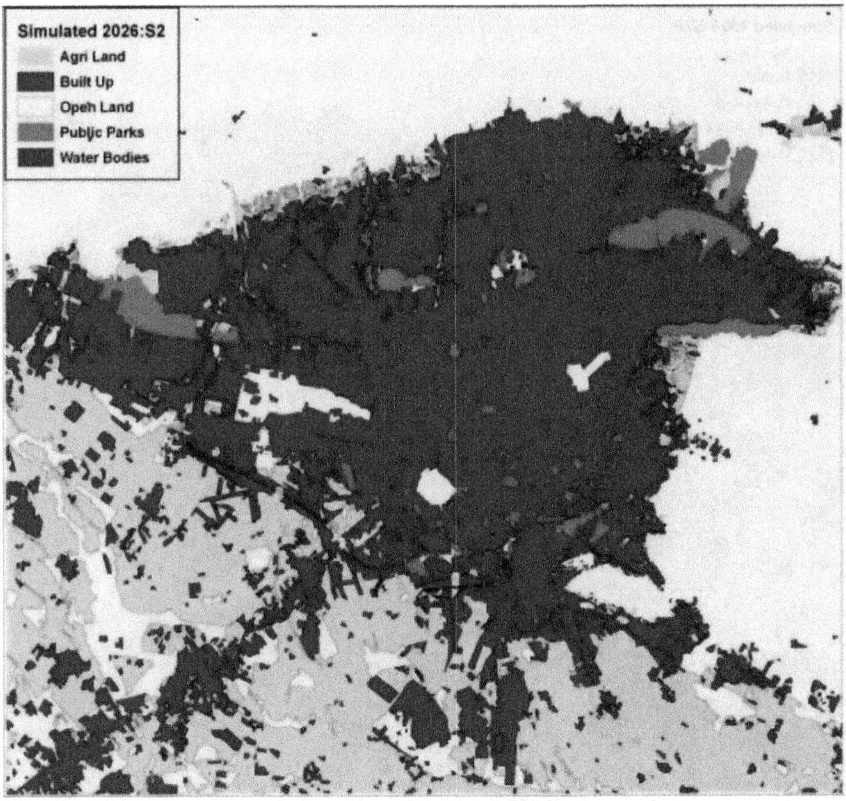

Fig. 7.12 Simulated map of 2026 by ABM simulation based on scenario 1

As was shown in the previous sections, the aim of MAS was to consider the internal and external interactions between autonomous agents and the overall organisation. This research has focused on various types of land cover change modelling and has compared the results to assess the strengths and weaknesses of each particular model. This research has also realised the drivers of change, as well as which sort of variables has more influence in this study area. It has also dealt with how to predict their behaviour within a given period. The agent-based modelling approach that has been carried out is flexible enough to be suitable for this research (Valbuena et al. 2008).

There is a large and ever increasing amount of research about using "*bottom-up*" techniques, such as cellular automata (CA) and other customised methods to simulate urban areas. The major problem in using the CA models is

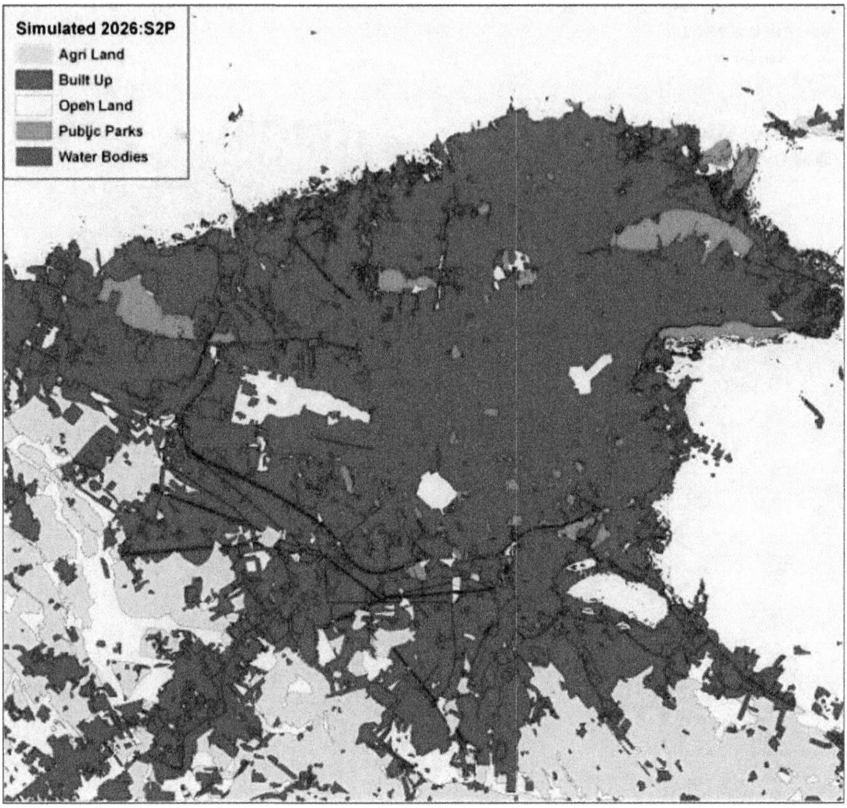

Fig. 7.13 Simulated map of 2026 by ABM simulation based on scenario 2

to integrate the human, social and economic factors to incorporate in the simulation. Table 7.2 summarises the implemented models and deals with the strengths and weaknesses of each particular model. More discussions and conclusions will be presented in the next chapter.

7.10 Summary

This chapter began with an introduction which presented the methodology of analysing the obtained results. The final simulated land use maps and their certainty were assessed. This assured us that the utilised maps are validated to input the designated models. Thereafter, the findings through the traditional techniques were discussed and the validity of those models was tested. Each model was statistically verified and the multi-agent model was evaluated step by step. Finally, the developed ABM system results were compared against the traditional

Table 7.2 Comparison of the executed land use change modelling models

Model	Required variables	Strength	Weakness
CA model	– Built-up cells	– Allocation starts from the nearest cells to the developed cells – A variety of rules can be coded in the model	– Does not consider human decisions that impact on the spread of built-up areas – Does not yet include bio-physical data – One land category can be modelled
Markov chain model	– Land use map	– Number of land categories unimportant – Results in more accurate quantity of change value – Transition probability maps are produced	– Non geospatial results – Does not consider any variable, must be integrated with geospatial models (e.g. CA, logistic regression)
CA-Markov model	– Land use map – No ancillary data	– Simulated map can be multi-categorical – Benefits from both CA and Markov models	– Does not take any variables into action – Simulated map would have unreal edges which do not match with reality
Hybrid-logistic regression model	CBD, demography, nearby cities, northing & easting coordinates, population density, residential area, buildings, farming lands, steams, DEM, interchange network, slope, open lands, parks, roads	– Probability of each cell is calculated –The produced probability surface allows the allocation of change based on any scenario	– Does not produce simulated maps – An extra technique must be carried out to produce simulated maps
Agent-based model	Elevation, slope, medical services, metro stations, disposal areas, orchards, sport centres, road networks, recreation points, commercial centres, railways, land price, housing rent, streams, airports, military facilities, power facilities	– Considers every available data – Has a geospatial structure – Weights each variable separately – Considers socioeconomic data – Integrates a CA function to allocate changes	– Requires a lot of data – Needs coding to consider agents' behaviour

approaches, which facilitated the general conclusion over the employed methods. In Chap. 8, a final conclusion of the results will be presented and the direction of future works will be drawn. Moreover, the assumed objectives of this thesis will be evaluated.

References

Li X, Liu X (2007) Defining agents' behaviors to simulate complex residential development using multicriteria evaluation. J Environl Manag 85(4):1063–1075

Maguire D, Batty M, Goodchild M (2005) GIS, spatial analysis, and modeling. Esri Press

Mousivand AJ, Alimohammadi Sarab A, Shayan S (2007) A new approach of predicting land use and land cover changes by satellite imagery and Markov chain model (case study: Tehran). MSc thesis, Tarbiat Modares University, Tehran

Oluseyi OF (2006) urban land use change analysis of a traditional city from remote sensing data: the case of ibadan metropolitan area, Nigeria. IDOSI Publications 1(1):42–64

Pontius RG Jr, Malanson J (2005) Comparison of the structure and accuracy of two land change models. Int J Geogr Inf Sci 19(2):243–265

Valbuena D, Verburg PH, Bregt AK (2008) A method to define a typology for agent-based analysis in regional land-use research. Agric Ecosyst Environ 128(1–2):27–36

Chapter 8
Conclusions and Recommendations

8.1 General Discussion

This thesis is compiled in eight principal chapters. Chapter 1 reviews the fundamentals of this research that were supposed to deal within this project (e.g. problem statement, research questions, research objectives and research approach). Chapter 2 presents a literature review of previous research work, and also deals with the theoretical background of related fundamentals.

Chapter 3 depicts a brief description of the selected study area, which is Tehran, the capital of Iran. This chapter includes more details of geographical characteristics, as well as the socio-economic conditions in the study area.

Chapter 4 explains the utilised materials and data for this project. A temporal mapping of land use changes is prepared and visualised. The trend of land use changes is also discussed within this chapter.

Chapter 5 clarifies various common approaches, which are popularly used (i.e. CA, the Markov model, the CA-Markov model, and hybrid logistic regression). These models were theoretically depicted and implemented to achieve the strengths and weaknesses of each particular model. These traditional models and their theoretical backgrounds are useful to find out the way to assemble the multi-agent model. Additionally, certain weaknesses of these traditional techniques are addressed, based on a lack of dynamics.

Chapter 6 depicts the assumed agent-based model. It begins with the classification of effective agents (i.e. resident, developer and government agents). The possible behaviours of each particular agent is described and modelled. The methodology of designing this ABM is depicted in detail, after which the predesigned agents are combined.

Chapter 7 assembles all the outcomes and analyses the achieved results, and the results of each stage are described and verified. A validation process of all models is then discussed and validated, and the models are compared with each other. Ultimately, the ABM model is compared against other alternative methods.

J. Jokar Arsanjani, *Dynamic Land-Use/Cover Change Simulation: Geosimulation and Multi Agent-Based Modelling*, Springer Theses,
DOI: 10.1007/978-3-642-23705-8_8,

This chapter clarifies the conclusions and the author's recommendations; thereafter, the suggested direction of future work will be explained, along with the limitations of this thesis.

8.1.1 Strengths and Weaknesses of Each Particular Model

Cellular automata models basically have the strength and capacity to start the allocation process of change from the nearest neighbour cells to the urban areas and might satisfy some land change modelling models. According to Tobler's first law of geography:

> Everything is related to everything else, but near things are more related than distant things (Johnston 2000).

Thus, urban sprawl most likely will take place alongside previous built-up areas. In our model nearby pixels to urban areas have more probability to be developed by individuals; however, it is easily observed via the temporal change mapping discussed previously.

The implemented cellular automata model was not integrated with any environmental and demographical parameters. Although the simulated maps were validated, integration of this model with other relevant parameters could increase the accuracy of this model. This makes this model very rigid and the modeller has to stop this process at a certain iteration number.

The Markov chain model is also another modelling approach. In fact, it does not produce any geographically explicit outputs. This method also does not consider any environmental and socio-economic variables. It predicts the quantity of change based on mathematical estimation and previous state of input land use maps.

The cellular automata Markov model is almost a new approach, which integrates both CA and Markov chain models. In fact, it retrieves the quantity of change from the Markov chain model and spatializes that through a cellular automata procedure. Nonetheless, it has to be stopped at certain iteration number. The number of iterations in this approach is large, and, therefore, is a hugely time consuming approach in comparison with CA or other approaches.

Logistic regression has this capability to be integrated with other techniques. In order to predict changes it creates a probability surface. This method considers environmental and socio-economic variables to produce a probability surface of change. Thereafter, this model was integrated with a Markov model to allocate the amount of change. This probability surface identifies the probability of change for each particular cell.

Multi-agent based modelling has the strength to integrate environmental and socio-economic variables and considers each particular agent which impacts on the system. Moreover, this approach by means of a "bottom–up" scenario considers possible variables associated with each single agent. This technique considers each agent's behaviour internally. Furthermore, ABM finalises all agents' interactions by combining all agents to find the location of future changes. The previous types of approaches are not able to reflect exactly the complex interactions between individuals' preferences and the environment. The ABM technique can be used to simulate complex built-up development which involves a variety of residents in shaping urban morphology. It is crucial to associate appropriate agents' behaviour in a more consistent way. This means that individuals are the main actors in this methodology.

In this research, the characteristics of agents were defined in a spatially explicit way to reflect various decision-making behaviours. It was aimed to avoid any use of ABM environments. Agents were classified into the most efficient ones according to available data. Furthermore, a weighting system was carried out over the available data to assign appropriate weights, which was helpful in considering their importance.

One advantage in particular of the ABM, is that it does not consider the nearest cells as the priority of change. There was a general lack of detailed social and economic data at a fine spatial scale, which encountered some difficulties in defining agents' properties in a more precise way.

8.1.2 Uncertainty Analysis

According to Crosetto and Tarantola (2001):

> Uncertainty analysis allows the analyst to assess the uncertainty associated with the model output as the result of the propagation through the model of errors in input data, and uncertainty in the model itself (e.g. uncertainty in model parameters, structures, assumptions and specifications).

In this study, much data and some models were employed. This data is somewhat erroneous, which needs to be depicted and analysed. For instance, the socio-economic data for the outskirts of Tehran city was not as reliable as those data for the city of Tehran. The difficulties in the preparation of the data did not allow us to reach accurate data; equally there are some doubts in the quality of the demography and land price layers. This could cause a lot of errors in the prediction process, besides which, models also have some weaknesses that have to be addressed. Each model is constructed on some hypothesis and manipulations which might not be necessarily true. In the ABM model, some formulae were used which might cause some errors. These two kinds of errors will be illustrated in the following sections.

8.1.3 Model Limitations

Each particular model has its own limitation and constraints. The cellular automata model is designed to grow out, based on straight rules at certain iteration numbers. Individually it does not consider environmental conditions;

however, some applications have used this technique to integrate with environmental parameters and drive it manually. This might create some errors in the results. The designed ABM model was developed by the author of this thesis, and it is a prototype which needs to be tested for similar cases to correct its probable bugs.

8.1.4 Data Limitations

In this research, collecting the required data was an important matter and also a time consuming procedure that took place during a field work period. Meanwhile, some sort of data, regarding accuracy assessment, was gathered for three separate time periods.

Data individually were not reliable and the accuracy of data had to satisfy us; therefore, data checking through data mining, satellite images and hard copy maps was carried out to increase the reliability of the data. In fact, the accuracy of these data can be one source of errors. The time scales of the utilised data (i.e. 1986, 1996 and 2006) were not precisely the mentioned times and, thus, this can bring another source of error to the results and conclusions.

8.2 ABM Method versus Alternatives

In this research, it was aimed to examine the geosimulation approach in contrast with other prevalent methods. As we discussed previously, regarding ABM's advantages and strengths, this approach takes into consideration key agents in any application and considers all possible actions arising from agents' behaviours. Nevertheless, awareness of each individual agent needs a comprehensive knowledge about those agents as well as qualified data at fine resolution. Although for this research, the provision of the required data at finer resolution was not feasible due to some difficulties for data sharing.

In fact, ABM benefits all prevalent methods and brings the advantages of the previous methods and makes it very organised.

8.3 Conclusions

Recent economic growth has created enormous problems in urban management matters, where urban expansion occurs without any concern for environmental impact. Rapid urbanisation and intensive land-use conflicts are occurring at an alarming rate in some developing countries. Therefore, the simulation and prediction of urban growth is imperative for urban planners to formulate sustainable development strategies. Besides, this helps them to predict land demands in order to provide enough infrastructures for the inhabitants.

Several approaches were performed to simulate land use change in the study area, such as CA, the Markov chain model, CA-Markov, logistic regression and multi-agent simulation. In this part of the chapter, a conclusion about the implemented methodologies, and also the obtained results, will be pointed out, to deal with optimum methods for LUCC modelling. Therefore, a summary of original conclusions will be presented later in this chapter.

We begin with the cellular automata approach and its strength and weakness. In fact, one difficulty with CA models is that in a pure CA model no human and social factors are incorporated in the simulation. Therefore, this model is not able to reflect the interactions between individuals and the environment. Some other customised CA models, however, do exist, which are able to combine environmental circumstances and their interactions with the system (e.g. SLEUTH model, CLUEs model). These models are built based on CA functionality and also incorporate some biophysical factors. We have concluded that the pure CA model, which does not take any environmental factors into account, is not a good location estimator for change prediction. However, the CA functionality, which starts the allocation process from the nearest cells, is the best way to allocate probable changes.

The Markov chain model is not a spatially explicit model to locate changes; however, it is a useful method to predict the amount of change. The predicted amount of change can be utilised in other land change modelling approaches to allocate it; for instance, in the frame of a CA model, or the logistic regression model. In this study, the output of the Markov model was used to allocate the predicted quantity of change through other methods (Kamusoko et al. 2009).

This research represents a significant contribution to land use modelling excluding integration of biophysical and socio-economic data into a spatially explicit Markov cellular automata land use simulation model. The strength of this model is to predict the change map for all existing land categories. In other words, the input and output files are categorical and do not have to be necessarily binary.

However, the weaknesses of this model are considerable. This model does not consider any environmental and socio-demographic situations. As a matter of fact, this model benefits from the output of the Markov chain model and, by means of a CA function, allocates the quantity of change. Moreover, this model needs to run for a large number of iterations, which takes much time and computational resources. Furthermore, this model allocates changes from the nearest cells to the urban cells.

A spatially explicit type of logistic regression modelling was implemented to discover and improve our understanding of the demographic, economic and biophysical circumstances that have driven land use change matter to discover the most probable sites of urban growth in the Tehran metropolitan area.

Urban expansion occurs essentially around existing urban cells, close to freeways and major roads, because of in situ infrastructure and excellent land market values. However, new areas located farther from existing urban areas might also have enough potential for development. A strong point of the logistic regression model is the ability to incorporate economic, demographic and environmental situations to produce a probability surface of potential urban growth. However, a CA function was implemented over the probability surface to allocate the anticipated amount of expansion in terms of the number of cells. The employed logistic regression model was spatially explicit.

This research proves that a logistic regression model has strengths relative to the previously mentioned models. The logistic regression model can not only comprise biophysical variables, such as SLEUTH (slope, land use, exclusion, urban extent, transportation, hill shade) in the CA model, but is also able to include a variety of demographic and econometric variables, which necessarily allows us to figure out human impacts in forming urban patterns.

Despite the LRM model's strengths, the logistic regression model suffers from the same constraints as the previous models in considering other issues, which may have an effect on urban growth (e.g. individual preferences for settlement, national and local development policies). Secondly, dissimilar to the CA model, the logistic regression model is not temporally explicit (Hu and Lo 2007). In fact, the output propensity surface can only point out where urban expansion might occur, but not specifically when it will happen. However, we utilised a CA function to overcome this weakness.

Multi-agent simulation can be designed to simulate complex urban development that involves a variety of agents playing significant roles. The definition of agents' behaviours process is an important task that has to be defined carefully. In this work, the most important agents which play considerable rules were taken into action (i.e. resident agents, developer agents, and government agents). Each agent was defined separately and, therefore, their own behaviours in choosing suitable cells for settlement were considered in a GIS environment. In fact, the multi-agent model has the capacity to consider all possible actions that can be made by the predefined agents. Furthermore, the ABM model does not start the allocation process from the nearest cell next to the city, meaning that the cells which have more propensity for development, become developed. However, in this study we combined the strengths of other methodologies to improve the quality of the predictor model. This approach benefits from cellular automata, Markov chain and logistic regression models.

This thesis has explored the potential for designing a geospatial multi-agent simulation model. Furthermore, it has successfully examined the proposed hypothesis and reached the assumed objectives. In fact, we benefited from a customized ABM model in a GIS environment to avoid using any ABM functionalities out of the GIS environment.

In this thesis, we accomplished the following objectives:

> To propose a generic method that can be followed to develop a multi-agent geosimulation system in GIS environments in various types of natural phenomena modelling:

Our research methodology reveals that GIS environments are capable of importing ABM functionalities and simulate natural phenomena within the current

GIS environments. Moreover, our approach is intended to be generic to integrate the strengths of the existing LUCC modelling methods and design an ABM system for predicting a system's behaviour.

To design an agent-based modelling prototype based on geographic data, GIS functions and promote the capability of GIS environments functionality for this matter:

In this thesis, we implemented several models that are being used for LUCC modelling. For instance, the cellular automata model, the Markov chain model, the cellular automata Markov model, the spatial logistic regression model and, finally, by combing the useful behaviour of each particular model, an initial version of multi-agent model was developed to simulate urban sprawl in the study area. This multi-agent model was designed according to the behaviour of three primary agents (residents, government and developer) and their interactions.

To propose an analysis technique to examine the results arising from the geosimulation performance in comparison with other methodologies, such as CA, Markov chains and hybrid models:

In this research, we studied different accuracy assessment techniques. It was illustrated that the ROC model is a better way to measure the accuracy of the outputs of the respective models, in terms of accuracy of quantity and accuracy of location and overall accuracy. However, some weaknesses still exist in the validation of outcomes.

To examine, assess and evaluate the existing software and toolkits that have been proposed to create simulation environment and their flexibility and compatibility due to importing geospatial data:

We started this work with testing other existing ABM toolkits such as Anylogic, Repast, agent-based extension, and each one had its own weakness to import GIS data and other spatial variables into the system. Then, we believed the best way was to code the ABM system behaviours into a GIS environment (i.e. ESRI ArcGIS) by means of its programming environment (e.g. Python). Therefore, this system is completely independent from the aforementioned toolkits' functions.

To consider the possibility of integrating GIS functions with ABM functions in a GIS environment and segregate geosimulation from the ABM environments:

In this thesis, a prototype model was developed without using ABM environments, but the GIS environments do not comprise all ABM functions, and it has to be coded by programming modules integrated in GIS software (e.g. Python).

To predict possible future changes within a particular period through the customised scenario:

The developed ABM system is able to predict upcoming changes at any requested time and updated data can be imported into the system in order to improve the quality and accuracy of outcomes. The system settings are available to modify the system and the data.

8.4 Directions for Future Works

The intention of this research was to explore the potential of the GIS environments to import ABM functionalities and create spatially explicit agent-based models which reflect geometric detail directly in the simulation process in order to simulate environmental phenomena.

As future tasks, it can be recommended:

- To gather and use detailed data in order to input into the geosimulation prototype and then take all agents' behaviours into account and obtain better outcomes. In other words, by using fine scale data, it is possible to reach better results.
- To use this simulation prototype and develop it by means of finer scale data to be used for land use change modelling and urban expansion issues.
- To follow up this type of agent-based system to develop new prototypes for other geospatial phenomena (e.g. hydrological modelling, human-related decision-making systems).
- To design a national project to simulate all possible behaviours of the existing agents and utilise the data of the entire metropolitan area, which can help land managers in better land administration.

8.5 Limitations of the Present Study

As mentioned in Chap. 5, we confronted a number of problems in the data gathering process, software restrictions for model development, and also weaknesses of model validation techniques. In fact, the socio-economic data specified, in the extent of study area, was not precise. The difficulties arising from the data gathering process did not allow us to acquire accurate demography and land price data. Another concern for this thesis was the limitations over finding appropriate models. In fact, the existing models have their own limitations and constraints.

Moreover, in this research all required data were not available, and the collection of the required data was also a time consuming procedure. Some of the utilised data individually were not completely reliable and the accuracy of data had to satisfy us; therefore, data checking through data mining, satellite images and hard copy maps was carried out to increase the reliability of the data. Indeed, the accuracy of these data can, in fact, be one of the sources of error. The time periods of the utilised data (i.e. 1986, 1996 and 2006) were not exactly the indicated time and, thus, can create error in the results and conclusions.

8.6 Original Guidelines in the Contributions of the Thesis

This dissertation has investigated a customised approach in order to simulate land use change. This approach is able to find potential cells for development and allocate them from the most probable to the least probable. In fact, the structure of cities is also an important matter to consider before developing any model in this regard. Indeed, it is not strictly correct to propose the recommendation of implementing agent-based modelling for any land change case. In other words, the structure of cities plays a critical role in selecting an appropriate methodology, i.e. it is better to have knowledge of temporal changes and analyse it by the pattern of local circumstances than to fit a model to the local conditions. Knowledge of a city's structure, as well as temporal assessment of development, are significant aspects to prescribe a methodology.

8.7 Summary

The last chapter presented a general discussion about this thesis and the tasks which have been carried out herein, followed by a discussion of the advantages and disadvantages of each model. The limitations arising from models, data and methodologies were depicted. The original guidelines which we revealed in this research were explained in detail, and suggested directions of future works in this field duly clarified.

References

Crosetto M, Tarantola S (2001) Uncertainty and sensitivity analysis: tools for GIS-based model implementation. Int J Geogr Inf Sci 15(5):415–437

Hu Z, Lo C (2007) Modeling urban growth in Atlanta using logistic regression. Comput, Environ Urban Syst 31(6):667–688

Johnston RJ (2000) The dictionary of human geography. Wiley, New York

Kamusoko C, Aniya M, Adi B, Manjoro M (2009) Rural sustainability under threat in Zimbabwe—simulation of future land use/cover changes in the Bindura district based on the Markov—cellular automata model. Appl Geogr 29(3):435–447